Literature-Based
Math Activities
AN INTEGRATED APPROACH

ALISON ABROHMS

SCHOLASTIC
PROFESSIONAL BOOKS

New York • Toronto • London • Auckland • Sydney

Designed by Nancy Metcalf
Production by Intergraphics
Cover design by Vincent Ceci
Cover illustration by Mona Mark
Illustrations by Julie Bertram

ISBN 0-590-49201-2

Printed in the U.S.A.

To my
beautiful daughters,
Megan and Alyssa,
who each light up my world
Equally!

Contents

Continued

UNIT 4

Measurement, Money, and Time 63

UNIT 5

Fractions, Probability, and Graphing 89

Support Masters 101

Introduction

Counting backward, counting sets to 10, constructing a number line . . . sounds like a math book, right? But it's not. It's *Ten, Nine, Eight,* by Molly Bang (Greenwillow, 1983), a favorite children's book, and a perfect book for teaching number sense and numeration.

When you use literature to teach math, you give children chances to develop mathematical understandings in a natural, more meaningful context. And that helps children develop the skills they need to express themselves—mathematically and otherwise. This book is designed to help you make those meaningful math connections in your classroom and to give you a springboard for recognizing and developing opportunities for teaching math with more of your students' favorite stories.

USING THE BOOK

From folktales to photo stories, the titles featured throughout this book all have two things in common—their appeal for children and their natural connections to math concepts. Titles are grouped into five math units:
1. Number Sense and Numeration
2. Whole Number Concepts
3. Geometry, Patterns, and Spatial Sense
4. Measurement, Money, and Time
5. Fractions, Probability, and Graphing

To help you meet your curriculum needs, concepts covered in each unit correspond to the new math standards as outlined in *Curriculum and Evaluation Standards for School Mathematics* (National Council of Teachers of Mathematics, 1989). Problem-solving is incorporated throughout. Nine reproducible support masters are included at the back of the book—you'll find grid paper, a gameboard, place-value models, a number cube, bills and coins, pattern shapes, and three pages of counters—all designed to support the reading, math, and special cross-curricular activities.

Within every unit, individual book lessons include:
○ a recommended title,
○ a story synopsis,
○ suggestions for shared-reading experiences,
○ math activities, and
○ a blackline student activity sheet.

You'll find that many of the activities in this book suggest partner and group work. As your students work together, encourage them to share and discuss their reasoning. You'll promote a cooperative spirit and give your students plenty of opportunities to clarify their thinking.

ASSESSMENT

To assess your students' understanding of the concepts developed in these activities, listen to their interactions. Ask children to describe what they are doing, to predict outcomes, and to justify any solutions or outcomes. You're bound to hear a lot more than answers. You'll hear children talking about math in meaningful ways—using math-related terms and concepts as they express themselves, and making connections to other curriculum areas. But don't let this math excitement end with the activities in this book. Be on the lookout for natural math connections in your students' favorite stories. You're sure to discover many more ways to use literature with the math you teach every day.

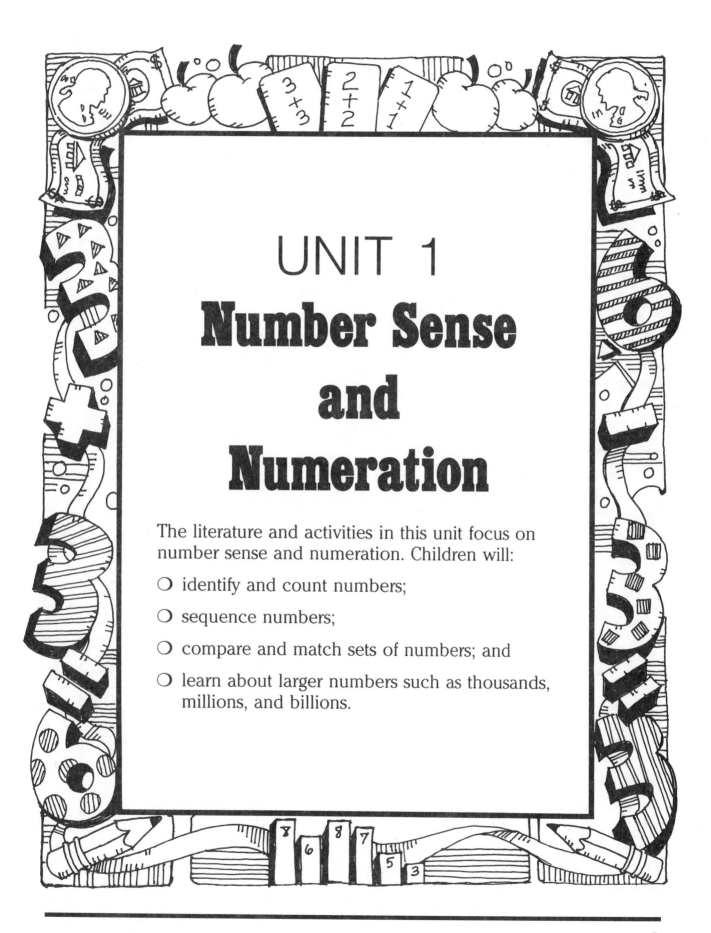

UNIT 1
Number Sense
and
Numeration

The literature and activities in this unit focus on number sense and numeration. Children will:

○ identify and count numbers;

○ sequence numbers;

○ compare and match sets of numbers; and

○ learn about larger numbers such as thousands, millions, and billions.

Over In the Meadow

by Olive A. Wadsworth
Scholastic, 1971; Penguin (Puffin), 1986

This classic Appalachian counting rhyme combines verse and brilliant illustrations to teach about animal mothers and their babies living in a meadow habitat.

Mathematical Concepts: Counting from 1 to 10, sequencing by size, graphing

SHARED READING

○ Discuss the cover illustration. Encourage students to recognize the meadow as a *habitat*—a place where animals find everything they need to live: food, air, water, and shelter.
○ List on the chalkboard or on chart paper the animals that children will read about in this story:

turtle	bluebird	honeybee	cricket	frog
fish	muskrat	crow	lizard	firefly

Ask children to share questions and information about any of these animals. As you read the story aloud, encourage children to predict each succeeding number once the counting pattern has been established.
○ Play *Over in the Meadow,* sung by Raffi, as you show the illustrations.

MATH ACTIVITIES

Ordering by Size

Give children Activity Sheet 1, scissors, paste, and crayons. Talk about the relative sizes of each animal shown. If possible, display other pictures of all or some of these animals. Ask children to cut out the ten animal pictures, sequence them by size in the boxes at the top of the page, paste the pictures in place, then label each picture in sequence from 1 through 10. A suggested sequence is shown, but children's interpretations may vary.

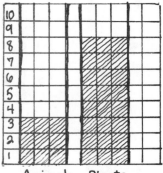

Class Graph

Make a class graph showing the number of living animals and the number of living plants in the classroom or school. When the graph is complete, count aloud the items recorded on the graph. Give each child a piece of grid paper to take home (Support Master 1). Help students label the grid paper as shown. Ask them to record the number of living things—animals and plants—in their homes.

Animals   Plants

ACROSS THE CURRICULUM

Science/Art

Have each child paint a picture of his or her favorite part of a meadow. Have children put their pictures together on a large piece of butcher paper to make a meadow mural.

Language Arts

Ask children how their homes are like animal homes in a meadow? How are their homes different? Have each child write or dictate a story about what makes his or her home a habitat.

Name _____

Over In the Meadow

lizard	crow	firefly	muskrat	honeybee
_____	_____	_____	_____	_____
cricket	turtle	bluebird	fish	frog
_____	_____	_____	_____	_____

Moja Means One: A Swahili Counting Book

by Muriel Feelings
Penguin (Dial), 1971

The language and culture of East Africa are introduced in this Swahili counting book. Scenes depicting the natural landscape and village life illustrate each Swahili number-word from one to ten.

Mathematical Concept: Counting sets of objects to 10

SHARED READING

❍ Ask children what they think the child in the cover illustration is doing. Read the story aloud then ask the same question.

❍ Give pairs of children ten counters (if you like, use those on Activity Sheets 7, 8, or 9). Reread the story. Have children say each Swahili word with you and show the corresponding number of counters.

MATH ACTIVITIES

Count the Drumbeats

Ask children to pretend that their desks or tables are drums. Have them practice counting to three in Swahili as they beat their drums: *moja, mbili, tatu; moja, mbili, tatu . . . .* Then, using a covered coffee can or oat box as a drum, have volunteers take turns beating once, twice, or three times. Ask the class to name the number of beats in Swahili.

Swahili Number Concentration

Give children copies of Activity Sheet 2, crayons, and scissors. Have them draw sets of objects to represent the numbers on the cards, then cut the cards apart. To play, have pairs of children arrange their combined cards in rows, face down on a table. The first player turns over two cards he or she thinks will match, and reads each Swahili number. If the cards match, the player keeps the cards. If not, the cards are returned face down. Play continues until all matches are found.

A Counting Game

Give pairs of children the bottom of an egg carton with two egg sections removed, (so there are ten cups), ten large seeds or other counters, and their Swahili number cards from Activity Sheet 2. In turn, one child displays a number card as the other child puts that many counters in the egg carton, one counter to a hole.

ACROSS THE CURRICULUM

Social Studies/Math

Invite children to locate Africa on a world map or globe. How far away do they think the closest part of Africa is to their state? Use a mileage key to calculate the distance.

Science/Art

Challenge children to create one of the animals pictured in the book using materials such as paints, crayons, yarn, and tissue paper.

Moja Means One:
A Swahili Counting Book

1 moja	2 mbili
3 tatu	4 nne
5 tano	6 sita
7 saba	8 nane
9 tisa	10 kumi

Count on Your Fingers African Style

by Claudia Zaslavsky
Crowell, 1980

Through rich illustration and vibrant prose, children are transported to a bustling African marketplace. Since English is not spoken at the marketplace, the children are encouraged to communicate their requests for three oranges using finger counting. As they travel around the marketplace, children explore traditional finger counting methods used by people who speak many different languages to communicate as they come together to buy and trade wares.

Mathematical Concepts: Counting, adding and subtracting, comparing

SHARED READING

❍ Ask children to describe how young children use their fingers to tell how old they are. Encourage children to think of other situations in which people show numbers using their fingers. As you read the story, allow children a few moments to imitate the finger counting methods as they are introduced.

❍ Reread the story and ask volunteers to write number sentences on the chalkboard that correspond to some of the finger counting methods illustrated in the story.

MATH ACTIVITIES

Guess My Number

Give pairs of children a copy of Support Master 2 (Gameboard), markers (such as coins), and a copy of Activity Sheet 3. Have students cut apart the playing cards on the activity sheet and place the cards in a pile face down. To play, one child picks up a card and demonstrates the hand position. The other child states the number being modeled. If the number is correct, that child moves the corresponding number of spaces. Children take turns modeling numbers until one player reaches the finish.

What Are the Facts?

Show children the finger counting indicated at the right. Ask volunteers to write addition and subtraction sentences to describe the finger counting on the chalkboard. Continue by letting children model other examples of finger counting, then call on volunteers to write the number sentences.

$$5 + 4 = 9$$
$$10 - 1 = 9$$

Comparison Shopping

Remind children that people in the story were trying to get the best prices as they shopped. Then provide small groups of children with two supermarket advertisements from a local newspaper. Ask them to compare the prices of three items found in each ad, such as a dozen eggs, a gallon of milk, and a head of lettuce. Ask children to share their findings with the class and tell who offers the best price for each item. Can the class predict, based on everyone's findings, which store might have the best prices overall?

Count on Your Fingers
African Style

Below are two ways to count to 8 African style.

A 1	A 2	A 3	A 4
A 5	A 6	A 7	A 8
B 1	B 2	B 3	B 4
B 5	B 6	B 7	B 8

Ten in a Bed

adapted by Mary Rees
Little, Brown, 1988

"There were ten in the bed." A delightful counting rhyme is repeated as the littlest child pushes all her friends, one at a time, out of the bed. Nine unhappy friends return with their own plan.

Mathematical Concept: Counting backwards from 10 to 1

SHARED READING

○ Show children the illustration on the cover and together count all the children in bed. Let children imagine what it might be like to have ten sleeping in their beds. Read the story aloud. As you say each number, have children hold up the corresponding number of fingers.

○ Give children copies of Activity Sheet 4 and scissors. Have children cut out the cards, then paste the corresponding number and dot cards back to back. As you reread the story, ask children to predict the number on the following page by holding up the card that shows that number.

MATH ACTIVITIES

You Tell the Story

Let groups of ten children experiment with retelling the story. First, have them each take a number card (from 1 to 10) and act out the events by holding up the correct cards as the numbers come up in the story. Next, have them take turns telling the story as they model the events with linking cubes. Just for fun, have children try to tell the story backwards, this time going from one in the bed to ten.

Scrambled Numbers

Let children practice arranging numbers in descending and ascending order. First, give groups of ten children a set of the number cards. Have them each choose a card then "shuffle" themselves around. Have children arrange themselves in ascending order, shuffle themselves up again then try descending order.

Counting With Cubes

Ask small groups of children to arrange their number cards in sequence from 10 to 1 on a desk or table. Provide each group with 55 linking cubes. Let children arrange the cubes to correspond with the number of cards as shown. Ask them to describe how the cube sets change with each counting number in the sequence. (Children may respond that there is one less cube each time.) Ask group members to take turns counting backwards orally. You may want to extend the sequence beyond 10.

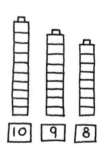

ACROSS THE CURRICULUM

Language Arts

Count down to a special event such as the 100th day of school, the number of days until a class trip, or the number of days until the next holiday with a class story or poem. Each day, add the part of the story or poem that corresponds to that day. Display your writing work in progress on a classroom wall, bulletin board, or even in the hallway outside the classroom door.

Ten in a Bed

1	2	3	4
5	6	7	8
9	10	●	●●

Activity Sheet 4

17

Ten, Nine, Eight

by Molly Bang
Greenwillow, 1983

Soft, glowing illustrations of a child's bedroom provide the setting for a visit with a father and young child who is preparing for bed. Together they count sets of objects around the room in a quiet and loving bedtime routine—from "10 small toes all washed and warm" to "1 big girl all ready for bed."

Mathematical Concept: Counting backwards from 10 to 1

SHARED READING

○ Ask children to share their bedtime routines. Tell children that this bedtime story counts down as a young girl gets ready for bed. As you read the story aloud, encourage children to predict the counting number on the next page.

○ After reading the story, ask children to describe how their bedtime routines are similar to that of the child in the story. On chart paper, make a list of routines that children have in common. Count them together.

MATH ACTIVITIES

Count Down

Put a twist on the hand game "odds-evens" to provide practice in counting backward from 10. Have students play the game in pairs. On the count of three, each student displays between zero and five (thumb included) fingers from one hand. Together, students count down from the combined number of fingers.

Game of Counting Sets

Form groups of 3–4 students each. Give each group a copy of Support Master 2 (Gameboard), Activity Sheet 5, scissors, and one game marker for each child. Have children cut apart their playing cards, shuffle them, and place them face down in a pile. In turn, each player draws a card, counts the number of objects aloud, then moves that many spaces on the gameboard. Play continues until all children reach finish.

Number Lines

Give each child a long strip of adding machine paper, paste, crayons, and a set of number cards for 1 to 10 (Activity Sheet 5). Let children color the sets of objects on each card, write the number in each set, then arrange and paste the cards in descending order on the paper strip to make their own "number lines."

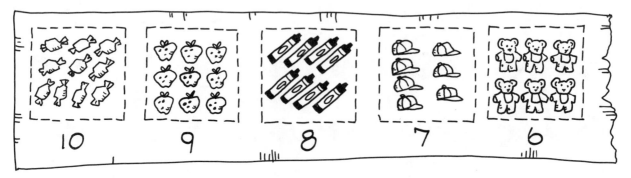

Ten, Nine, Eight

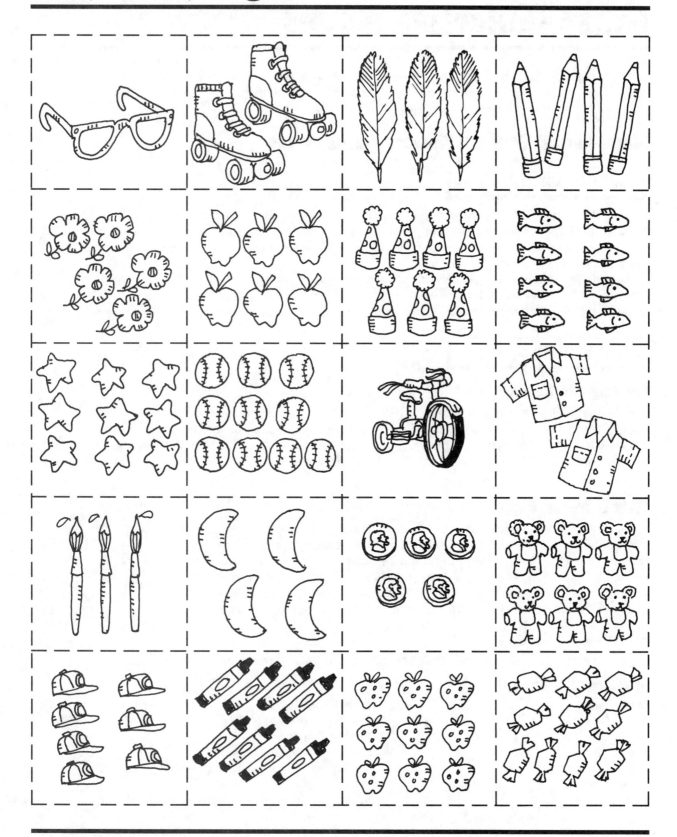

A Cache of Jewels and Other Collective Nouns

by Ruth Heller
Putnam, 1989

"A cache of jewels, a batch of bread, a school of fish, a gam of whales"—these and other collective nouns describe animals, plants, fruits, and other groups of objects. Children can count the beautifully illustrated objects on each page as they listen to the rhyming text.

Mathematical Concept: Numbers through 100

SHARED READING

○ Read the story aloud. Ask children to name some of the collective nouns they remember from the story. Record their answers on chart paper. Encourage children to name other collective nouns they know and include these on the list. Display this list and encourage children to add other collective nouns they learn about.

○ Reread the story and ask volunteers to estimate the quantity of each group of objects. Record estimates on the chalkboard. Let one child count the objects in each group. Help children compare the counts to the estimates, asking them to tell if estimates are high or low.

MATH ACTIVITIES

Creating and Naming Sets

Ask children to use their imaginations or explore the classroom to find a group of objects they would like to illustrate. Ask children to paint or color a realistic version of their group of objects. Help children write descriptions of their groups using collective nouns and giving the quantity, such as, "a gaggle of 10 geese." After paintings have dried, create a class book. Display the book in your classroom or school library.

Make-Your-Own Bugs

Remind children that insects have six legs. Ask children to help you name some insects. Then give children copies of Activity Sheet 6, scissors, paste, crayons, and construction paper. Ask them to cut out all 24 legs. Children should attach the legs to the insects, color or decorate the insects, and add their finished work to a table where the collection will be kept. Together choose a collective noun that describes the set, then write this on an oaktag strip and place it on the table.

A "Batch" of Cookies

Bake a "batch" of cookies with the class. Let children assist in measuring all the ingredients.

Sugarless Banana Cookies

Ingredients:
4 large bananas
2 cups oats
1/3 cup oil
1 tablespoon vanilla
3/4 teaspoon salt
1/2 cup raisins

Directions:
1. Preheat oven to 350 degrees.
2. Mash bananas.
3. Stir in remaining ingredients.
4. Place tablespoon-sized mounds on ungreased cookie sheets.
5. Bake 15–20 minutes. Makes 20 to 24 cookies.

A Cache of Jewels and Other Collective Nouns

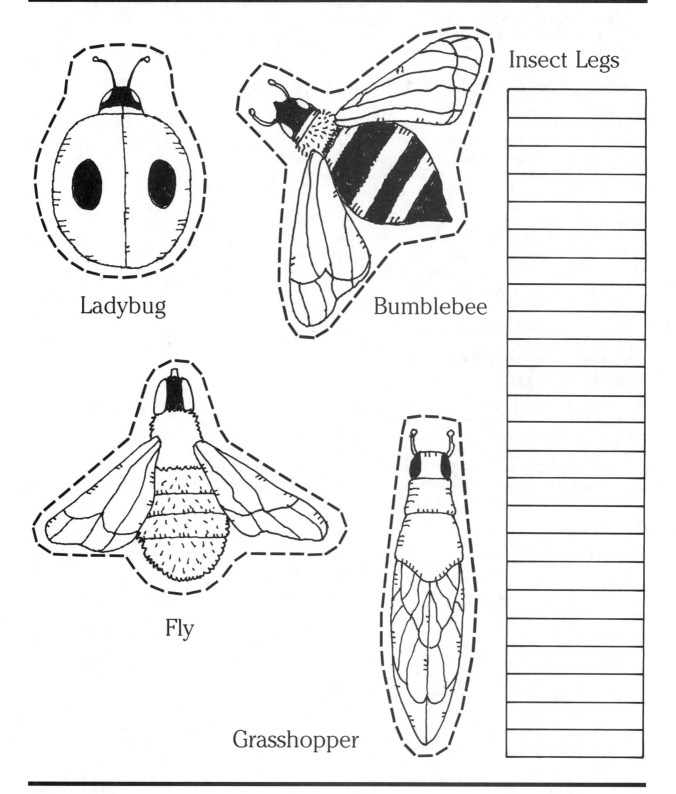

Ladybug

Bumblebee

Insect Legs

Fly

Grasshopper

The 329th Friend

by Marjorie Weinman Sharmat
Four Winds, 1979

Emery Raccoon woke up one morning bored and unhappy with his own company. He decides to invite some guests for lunch, 328 to be exact, hoping that just one will like him. He painstakingly set the tables and cooked the meal. But when his guests sit down to lunch, they pay no attention to him. That's when Emery Raccoon discovers who his best friend truly is. In this warm-hearted picture book, children will share with Emery the joys of self-discovery.

Mathematical Concepts: Numbers through 400

SHARED READING

○ Ask children to think about how many friends they have had at their homes at any one time. Then show them the cover illustration and introduce Emery Raccoon. Tell children that as they listen to the story, they'll learn what happens when Emery invites 328 friends to his home for lunch.

○ Engage children in a discussion about the qualities Emery discovers about himself. Ask children to think about which of the same qualities they like about themselves.

○ Give pairs of children place value blocks or Support Master 3 and scissors. To help children conceptualize the number 329, ask them to model 329 with place value models. Encourage children to imagine that each block represents one friend. Talk about where they might see groups with that many people.

MATH ACTIVITIES

Counting From 301

Tell children to pretend that Emery Raccoon counted all his guests as they came to his home. To help them practice counting, give groups of children 29 index cards or cards cut from construction paper on which the numbers 301–329 have been written (one per card). Ask children to work with their group members to arrange the number cards in order. When children have finished, lead them in reading the numbers in unison, having them point to each number as they say it. Remove some cards from each group so the numbers are not consecutive. Ask children to shuffle the cards, then arrange their numbers in order.

Number Game

Divide children into groups of three and give each group Activity Sheet 7. Have children cut apart the number grid, place value chart, and the ten-more chart. One child begins by tossing the counter onto the number grid and identifying the number it lands on. Another child writes the number in the place value chart. The third child writes the number that is ten more. Ask children to take turns at each role.

324	230	200	188	161
272	314	209	190	156
247	268	315	305	123

Hundreds	Tens	Ones
3	1	4

Number	Ten More
314	324

The 329th Friend

324	230	200	188	161
272	314	209	190	156
247	268	315	305	123

Hundreds	Tens	Ones

Number	Ten More

Activity Sheet 7

23

How Much Is a Million?

by David M. Schwartz
Scholastic, 1986

Marvel the Mathematical Magician guides children into the world of large numbers. His magic will help children explore the enormity of the tallest buildings and the highest mountains, how long it takes to count to one million, and how much space one million takes up.

Mathematical Concepts: Numbers through one trillion

SHARED READING

❍ Ask students to imagine how tall one million children would be if they climbed onto one another's shoulders. After some discussion, read the story aloud.

❍ Give each child a copy of Activity Sheet 8. Help children cut and paste the pieces to form a place value chart. As you reread the story, ask children to write each number mentioned in the story on the chart. (1,000,000; 23; 100; 70; 100,000; 7; 1,000,000,000; 95; 10; 200,000.) Note: children will need to write one trillion on a separate sheet of paper.

MATH ACTIVITIES

How Far Can You Count?

Seat children with a partner. Ask children to estimate how far they can count aloud in two minutes. Using a timer or the classroom clock, ask one child in each pair to begin quietly counting from one on your signal while the partner monitors the count. At the end of two minutes stop the count and ask partners to record how far they counted. Then let children switch roles as you time them again. Help children compare their counts to their estimates.

Place Value Game

Children can play this game in small groups. Each child will need Activity Sheet 8 and each group will need Support Master 4 (Number Cubes). Ask group members to shuffle their sets of number cards and place them face down. One child rolls the number cube. Each group member draws that many number cards and arranges the cards on his or her place value chart to form the greatest number possible.

millions	Ten-thousands	thousands	hundreds	tens	ones
8	7	6	4	3	0

ACROSS THE CURRICULUM

Science

Divide children into nine groups and assign each a planet. Ask each group to find the distance that planet is from the sun. Give each group a copy of Activity Sheet 8 on which to record their number. Let each group report their findings to the class.

Average Distances From the Sun (source: *The World Almanac*, Pharos Books, 1990)

Mercury	36,000,000	Mars	141,750,000	Uranus	1,764,500,000
Venus	67,250,000	Jupiter	483,800,000	Neptune	2,791,050,000
Earth	93,000,000	Saturn	887,950,000	Pluto	3,653,900,000

How Much Is a Million?

paste here	ten-thousands	thousands	hundreds	tens	ones
ten-billions	billions	hundred-millions	ten-millions	millions	hundred-thousands

0	1	2	3	4
5	6	7	8	9

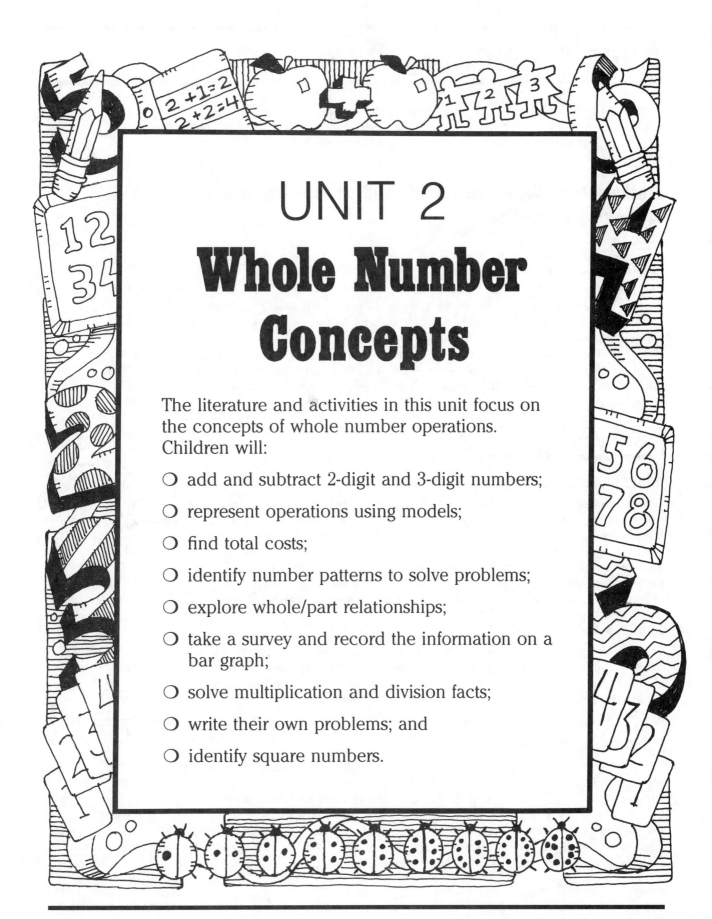

UNIT 2
Whole Number Concepts

The literature and activities in this unit focus on the concepts of whole number operations. Children will:

❍ add and subtract 2-digit and 3-digit numbers;

❍ represent operations using models;

❍ find total costs;

❍ identify number patterns to solve problems;

❍ explore whole/part relationships;

❍ take a survey and record the information on a bar graph;

❍ solve multiplication and division facts;

❍ write their own problems; and

❍ identify square numbers.

Anno's Counting House

by Mitsumasa Anno
Putnam, 1982

As we peek inside a house, we see ten little people with all their possessions. One by one the people move from one house to another, taking their possessions with them. Warm and captivating illustrations provide innumerable opportunities for counting, adding, and comparing groups of people.

Mathematical Concepts: Adding and subtracting facts to 10, sorting by attributes

SHARED READING

❍ Explain that ten little people live in a house (show front cover). One at a time, they move to another house (back cover). Before you read the story, let a volunteer count the number of people in the first house to establish that there are ten. As you read, let children count along to determine the number of people in each house.

❍ Cut out the 10 figures on Activity Sheet 9. Paste a piece of felt on the back of each and put the set of people on flannelboard. Use two large felt squares to represent the houses. As you share the book again, let volunteers model the movement of people between houses on the flannelboard, while the rest of the class counts the groups of people in each house.

MATH ACTIVITIES

Adding and Subtracting

Give each child 10 counters (if you like, use those on Support Master 7), two sheets of construction paper, and a piece of paper for recording. Have children draw a house on each piece of construction paper. Show a page from the book and have children place the counters accordingly in the two houses. Ask children to write addition and subtraction sentences about the model. Let children share their number sentences with the class. Finally, ask children to show their own families moving from one house to another then write number sentences about the pictures.

$$9 + 1 = 10$$
$$10 - 1 = 9$$

Spill the Beans

Give pairs of children a cup and 10 counters. One child spills all the counters onto a desk or table, then quickly gathers up as many of the counters as he or she chooses and hides them under the cup. The partner determines how many are hidden using the total number (10) and the counters remaining on the table. Children change roles and continue.

Patterns and Sorting

Show children the book again and encourage them to describe some things they notice about the people. For example, they may notice that some are boys and some are girls, some are wearing hats and some are not, some are in striped clothes and some are in dotted clothes, and so on. Give pairs of children Activity Sheet 9, crayons, paste, and craft sticks. Children cut out and color the ten figures, then paste a craft stick on the back of each. Bring children together and have them display their puppets in the center of the group. Ask volunteers to suggest ways to sort the puppets. Let children sort by some of the attributes suggested.

Anno's Counting House

10 for Dinner

by Jo Ellen Bogart,
(Scholastic, 1989)

On her birthday, Margo invites 10 friends for dinner. The party is a great success with food, arts and crafts, songs, and games. All the guests enjoy the activities Margo plans, except for one child who chooses his own expressive way to enjoy the festivities. Children will enjoy adding along with this whimsical counting book.

Mathematical Concepts: Representing numbers, addition

SHARED READING

○ Encourage children to describe the ideal birthday party. Display the book cover and tell children that Margo is celebrating her birthday by inviting 10 friends for dinner. As they listen to the story, ask children to watch the mischievous child in the red suit, then decide whether they would like him for a friend and explain why.

○ Give each child 10 linking cubes or counters. Reread the story and ask children to link cubes or arrange counters to represent the number of friends participating in each activity.

MATH ACTIVITIES

How Many Guests?

Let children work in pairs or small groups to write number sentences about the party activities described in the story. First, write the information about the party activities on index cards as shown. Give each pair or group one of the index cards, a number line, and counters. Have children use the counters and number line to show the total number of guests involved in the activity, then write an addition sentence about the information. Plan group time for students to share the pages that correspond to their activities and to show how they represented the addition.

I arrived at 3:45
5 guests came at 5:00
2 came at 5:10
2 came at 5:15

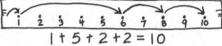

$1 + 5 + 2 + 2 = 10$

Play Party

Invite small groups of children to plan a make-believe party for ten guests. Give each group a copy of Activity Sheet 10 and sales circulars (groceries, toys, etc.) and have them circle and cut out items they want for their party and note the price of each on a separate sheet of paper. Have children work together to sort their party items according to the groups on the activity sheet (foods, beverages, party favors, decorations). If any category is blank, students should return to the sales circulars to find appropriate items. Have each group finalize decisions on party items, paste the pictures in place, then add up the cost of the party supplies. Let groups share their party plans and the total cost of each.

ACROSS THE CURRICULUM

Health/Math

Prepare a class party snack. Ask students to estimate how many cups of fruit salad they can make if each child brings in one piece of fruit. Send home a note requesting that each child contribute one piece of fruit (or a handful or berries or grapes). Wash and cut up the fruit. Let children take turns measuring how much fruit salad they make. Invite ten friends to share your treat!

10 for Dinner

FOOD

BEVERAGES

Party Favors

DECORATIONS

The Boy who Was Followed Home

by Margaret Mahy
Dial (Penguin), 1986

Robert is the sort of boy that hippopotami will follow, and follow him they do! When the number of adoring hippos frolicking on the family's front lawn finally reaches twenty-seven, Robert's father decides things have gotten out of hand. An unconventional solution to the problem produces delightfully unexpected results!

Mathematical Concepts: Subtraction, number patterns

SHARED READING

○ Ask children to describe Mary's predicament in the nursery rhyme, "Mary Had a Little Lamb." Then ask children to share situations in which they have been followed by an animal.
○ As you read the story, have volunteers record on the chalkboard the number of hippos in Robert's front yard each day. Ask children to look at the list of numbers (1, 4, 9, 27, 43) and describe any relationships they notice. For example, the numbers are in order from least to greatest; more hippos come each day.

MATH ACTIVITIES

How Many More Hippos?

Give children the top portion of Activity Sheet 11, and place value blocks or Support Master 3 (Place Value Models). Let children work with a partner to find how many more hippos were in front of Robert's house on each count. Ask them to record the differences on the activity sheet. Then ask children to share with the class how they computed the differences.

Predicting How Many More

Tell children that when the giraffes started following Robert home, he began keeping track of how many he saw each day, and found a way to predict how many would follow him the next day. Give children the bottom section of Activity Sheet 11 and have them work in pairs to first find then continue the number pattern for the two weeks shown. Children can use their number pattern to answer questions such as, "How many giraffes on the first Friday?" "On the second Monday?" "On the second Wednesday?"

What's the Difference

Pairs of children can play this game: Partners clap hands in unison three times. In place of a fourth clap, each child holds up any number of fingers (0-10). Each partner tries to be the first to find and call out the difference between the number of fingers each holds up.

ACROSS THE CURRICULUM

Language Arts/Dramatic Arts

Ask children to write a story telling what the hippos said to each other about Robert after he took the witch's pill. Give children opportunities to give dramatic readings of their stories for the class.

The Boy who Was Followed Home

	Number of hippos	How many more hippos than yesterday?
Day 1		
Day 2		
Day 3		
Day 4		
Day 5		

Sunday	Monday	Tuesday	Wednesday	Thursday	Friday	Saturday
1 giraffe	4 giraffes	7 giraffes	10 giraffes	13 giraffes	___ giraffes	___ giraffes
___ giraffes	___ giraffes	___ giraffes	___ giraffes	___ giraffes	___ giraffes	___ giraffes

Activity Sheet 11

The Philharmonic Gets Dressed

by Karla Kuskin
HarperCollins, 1982

It is Friday evening and one hundred and five people are getting dressed to go to work. As we watch them get ready, we learn that ninety-two of the people are men and thirteen are women; forty-five men stand up to pull on their pants and forty-seven sit down; and eight women wear long black skirts with blouses or sweaters, four wear long black dresses, and one wears a black jumper over a black skirt. Chatty text and delightfully candid illustrations are sure to charm and tickle children.

Mathematical Concepts: Exploring whole/part relationships, adding and subtracting to three digits, exploring attributes

SHARED READING

❍ Introduce the story by reading the title to children and ask them to tell what the *Philharmonic* might be. Encourage children to look for clues in the cover illustration.

❍ Invite children to share stories about concerts they've attended. Ask them to describe what musicians wore, what instruments they played, and about how many musicians performed.

MATH ACTIVITIES

The Whole and the Parts

Have children copy the whole-part-part chart as shown. Reread the first few pages of the story to help them recall the number of people in the orchestra, (92 men and 13 women). Children can use place value blocks or Support Master 3 (Place Value Models) to help them identify the parts and the whole. Ask children to write the numbers in their charts, then write addition and subtraction sentences using the parts and the whole. Reread the rest of the story, stopping to let children model and write number sentences for other groups of numbers.

Whole 105	
Part 92	Part 13

$$92 + 13 = 105$$
$$105 - 92 = 13$$

Sock Survey

Arrange the class in two groups. In this activity, children take a survey of the types of socks their group members are wearing and record the results on Activity Sheet 12. Children can decide how they will survey their group members and work cooperatively to complete the graphs. Have each group present their results to the class. Combine each group's graphs to create a whole-class graph (on chart paper or the chalkboard). Let students compare their group results with those of the entire class.

ACROSS THE CURRICULUM

Music

Put together your own Philharmonic with instruments students create from materials around the room. Instruments students can make include:

❍ varying amounts of water poured into glasses that can be tapped with spoons,

❍ metal table or chair legs that can be gently struck with a pencil or pen,

❍ small, loose objects (such as pennies, checkers, or dried beans) placed inside a box that is then taped shut, and

❍ rubber bands stretched between objects of varying distances.

The Philharmonic Gets Dressed

Sock Survey

Color	1	2	3	4	5	6	7	8	9	10
White										
Black										
Blue										
Green										
Brown										
Red										
Patterned										
Other										

Type	1	2	3	4	5	6	7	8	9	10
Sweat Sock										
Ankle Sock										
Knee Highs										
Tights										
Other										

Mice Twice

by Joseph Low
Macmillan, 1986

Cat would love nothing more than to dine on a nice, tender mouse. When Cat invites Mouse to join him for dinner, Mouse cleverly asks if she may bring a friend. "Mice twice!" Cat thinks to himself with relish. Mouse will surely have children cheering her on as she matches wits with Cat in this hilarious Cat and Mouse tale.

Mathematical Concepts: Multiplication, exploring relative size

SHARED READING

○ Ask children to guess what the title, *Mice Twice,* might mean. As you read the story aloud, ask children to think about their guesses and decide if they were accurate.

○ Reread the story and ask children to listen for ways in which parts of the story *increase* as the story progresses. For example, the dinner guests become larger and more ferocious, the dinner time is later each day, and the food becomes more abundant and elegant. Pairs of children may enjoy acting out "twice as big" and "four times as ferocious" for the class.

MATH ACTIVITIES

Twice the Number of Mice

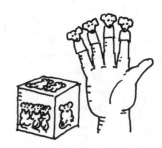

Give pairs of children Activity Sheet 13, scissors, crayons, and paste. Help them assemble the number cube and the finger puppets. To play, one partner tosses the cube, and the other partner shows twice the number of mice by placing the correct number of finger puppets on his or her hands. After children have played several rounds, ask them to tell what multiplication sentences they are modeling. ($1 \times 2 = 2$, $2 \times 2 = 4$, $3 \times 2 = 6$)

Cheese-Tasting Party

Remind children of the different types of cheeses that are served in the story. Tell them they will be sampling some different types of cheese. Cube a pound each of white and yellow cheese. Group children in fours. Provide each group a plate of mixed cube cheese and four small paper plates. Ask children to place 2 cubes of same colored cheese on their plates. Then have them place twice as many of the other color cheese on their plates. Tell children that you would like them to try each type of cheese and decide which they prefer. Show children's preferences on a graph.

Twice as Long

Give pairs of children rulers marked in inches. Ask one partner to choose a classroom object small enough to fit in one hand. Have the other partner find an object that he or she estimates to be twice as long. Have children trace around the objects on paper and record the length next to each shape.

Mice Twice

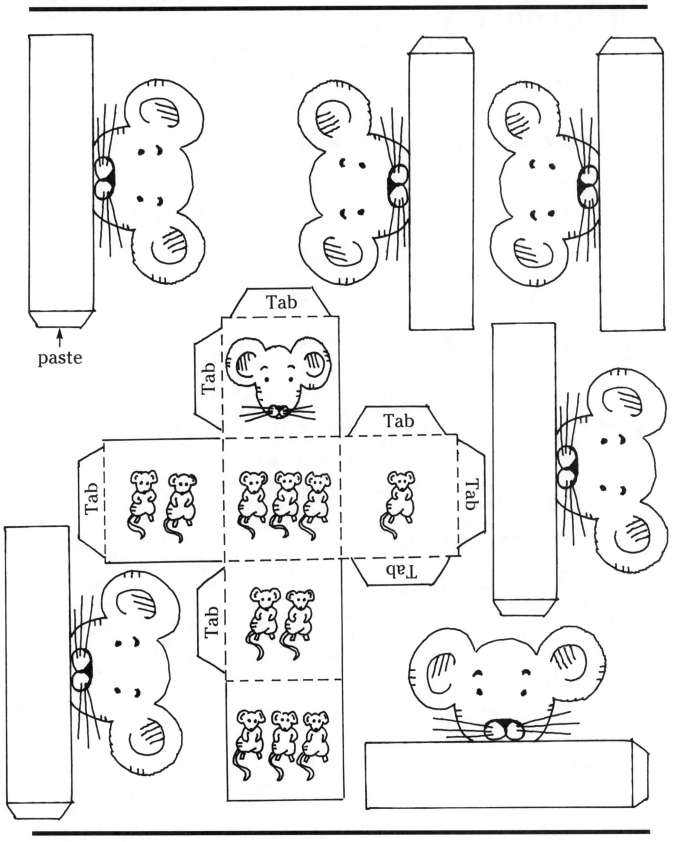

paste

Tab

Tab

Tab

Tab

Tab

Tab

17 Kings and 42 Elephants

by Margaret Mahy
Dial (Penguin), 1990

17 kings and 42 elephants go on a rollicking romp through the jungle, meeting all sorts of wild and wonderful creatures along the way. The hilarious tongue-twister verse and vibrant illustrations make this a jungle journey children will want to experience again and again.

Mathematical Concepts: Addition, subtraction, and multiplication

SHARED READING

○ Before reading the story, ask children to name animals that are used around the world for transporting people.

○ Reread the story and have volunteers count the number of animals that the kings and elephants meet along the way. Record the numbers on chart paper for children to use in "Live from the Jungle." (hippos, 4; tigers, 5; cranes, 4; gorillas, 3; pelicans, 8; peacocks, 2; flamingos, 6; birds, 9; baboons, 6)

MATH ACTIVITIES

Jungle Play

Have children dramatize scenes from the jungle that represent addition, subtraction, and multiplication problems. First, cut apart the activity cards on Activity Sheet 14. Have children take turns selecting a card at random, choosing enough classmates to play the parts of the animals represented on the cards, and dramatizing a mathematical problem. Students in the audience should try to solve the problem. For example, students dramatizing a card that pictures three gorillas and nine birds might do the following: three students moving like gorillas join nine students chirping like birds to represent the equation $3 + 9 = 12$. A card that simply pictures five tigers can be represented by five children growling like tigers to show $5 \times 1 = 5$.

Live From the Jungle

Pairs of children can pretend that two of the kings on the journey are actually a team of television reporters. Ask children to use the information saved on chart paper (from the *Shared Reading* section) to make a graph showing the various numbers of animals that were seen along the way. Children can use Support Master 1 (Grid Paper) to draw their graph. Then have them write a newscast using the information from the graph. Encourage children to include number comparisons using "more than" and "fewer than." Teams of reporters can take turns "broadcasting" their reports from the jungle.

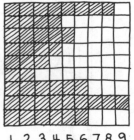

pelicans
hippos
tigers
cranes
gorillas
peacocks
flamingos
birds
baboons

1 2 3 4 5 6 7 8 9

ACROSS THE CURRICULUM

Language Arts

Have children work in cooperative groups to write and illustrate an epilogue to *17 Kings and 42 Elephants* in which they tell where the procession is going, and what happens when it gets there.

17 Kings and
42 Elephants

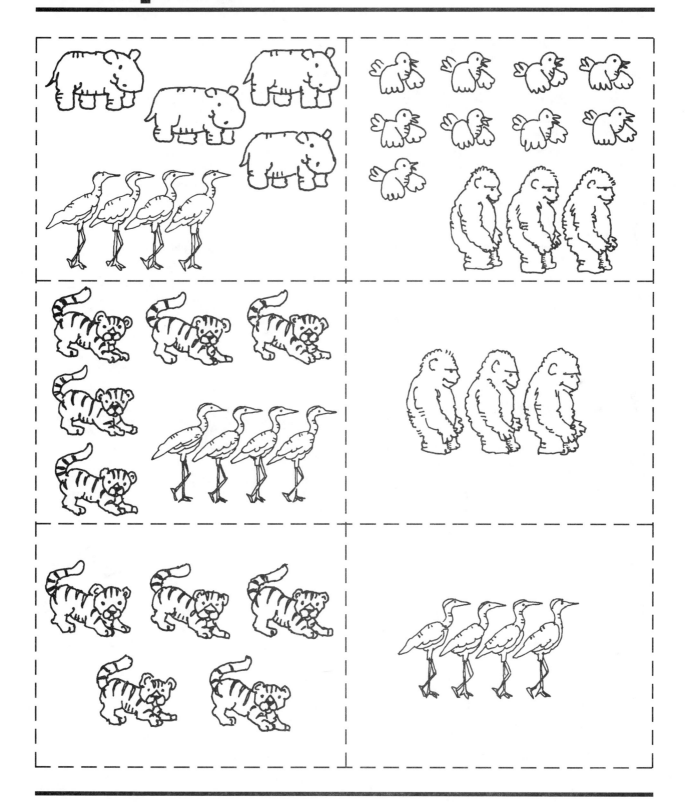

The Doorbell Rang

by Pat Hutchins
Greenwillow, 1986

Just when Sam and Victoria had determined their equal shares of Ma's freshly baked cookies, the doorbell rings. Ma invited two friends to share the cookies. "That's three each," figures Sam and Victoria. Then the doorbell rings again. As the scenario continues, the children recalculate the number of cookies the guests will receive as each new guest arrives, so the cookies are always shared equally. When each child has only one cookie . . . the doorbell rings again.

Mathematical Concept: Dividing by making equal shares

SHARED READING

○ After reading the story, ask students to describe what happens to the number of cookies each child will receive as each new guest arrives. Ask children to explain why Sam suggests waiting for Ma to open the door each time, before everyone eats their cookies.

○ Display pages that show one group at the table and group at the door. Ask volunteers to suggest a number sentence that describes each pictures.

○ Ask children to look again at the illustrations and describe all the ways that the kitchen scene changes to show the passing of time in the story. For example, the cat moves, the pot and kettle on the stove boil, and the number of muddy footprints on the floor increases as the story progresses.

○ Encourage children to identify patterns they see in the illustrations such as the checkered table cloth and floor, striped shirts, and borders around the plates and rug.

MATH ACTIVITIES

How Many Cookies Each?

Group children in fours. Give each group Activity Sheet 15, scissors, crayons, 12 counters to represent cookies, and 12 small paper plates or circles cut from paper. Have group members cut apart the cards on the activity sheet, then use the counters and the plates to model and solve the problems. Children can record their work by drawing the correct number of cookies on the plates shown. During a sharing time, encourage children to describe how they solved the problems.

ACROSS THE CURRICULUM

Language Arts/Math

Ask children to bring in a favorite family recipe for cookies. Have children copy their recipes on chart paper. Display the charts across the chalkboard tray and use them as springboards to counting, sorting, classifying, and other activities. Find out, for example:

○ Which recipe uses the most flour?

○ The least sugar?

○ How many recipes say to bake at 350 degrees?

○ If we made all of these recipes, how many dozen eggs would we need all together?

○ Which recipes call for nuts? for raisins? for chocolate chips?

　Bind the chart papers to make a Big Best Cookies Book. Let children take turns taking the book home to share with their families.

The Doorbell Rang

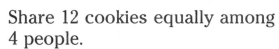

Share 12 cookies equally
between 2 people.
How many does each person get?

Share 12 cookies equally among
4 people.
How many does each person get?

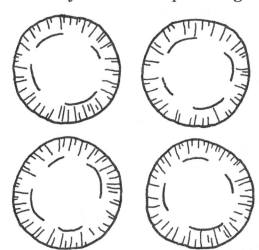

Share 12 cookies equally among
6 people.
How many does each person get?

Share 12 cookies equally among
12 people.
How many does each person get?

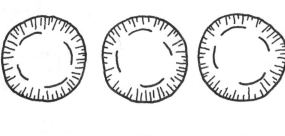

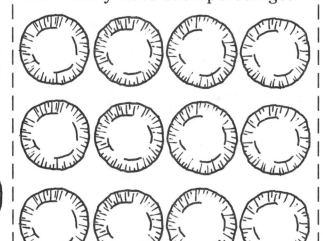

Bunches and Bunches of Bunnies

by Louise Mathews
Scholastic, 1991

Brightly colored, friendly bunnies hop around in a variety of activities from "planting seeds and pulling weeds" to "drinking punch and waiting for lunch." Each rhyming chant counts groups of bunnies and introduces a corresponding multiplication sentence.

Mathematical Concept: Modeling multiplication facts, square numbers

SHARED READING

○ Show children the cover illustration. Ask them to tell if the bunnies would be easier to count in a large group as shown here, or grouped so they could be counted by twos, threes, or fours. Encourage volunteers to justify their reasoning. As you read the story aloud, let children predict the multiplication sentence for each set of bunnies before you read it.

○ Give small groups of children 81 counters. Then as you reread the story, write each multiplication sentence on chart paper and ask children to model it with their counters.

MATH ACTIVITIES

Bunches and Bunches of Carrots

Provide pairs of children with several copies of Activity Sheet 16, paste, scissors, and paper to make a booklet. Assign each pair of children a multiplication table, such as the four table. Ask children to write one multiplication sentence on each page of their booklet, cut out and paste corresponding bunches of carrots to illustrate each sentence. Students can cut out the bunny at the bottom of each Activity Sheet and use it to create covers for their books. Have students staple their pages together then share their booklets with the class.

Multiplication Squares

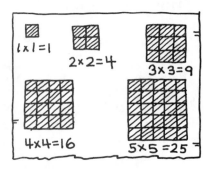

Give pairs of children Support Master 1 (Grid Paper), scissors, paste, crayons, and construction paper. Ask children to color and cut out a model for each multiplication sentence written on chart paper (from the *Shared Reading* section). Have children paste these models on construction paper, then write the corresponding multiplication sentence below the model, as shown. Call on volunteers to describe the shape of each model. (Children should notice that each model forms a square.)

More Multiplication Squares

Ask children to predict if the product of any number multiplied by itself will form a square. Then ask groups of children to each choose a number between 12 and 20, multiply the number by itself on a calculator, then make a model of the multiplication on grid paper. For larger numbers, children may need to cut out several sheets of grid paper and paste them on a bigger sheet of paper. Discuss why these numbers are called *square numbers*.

Bunches and Bunches of Bunnies

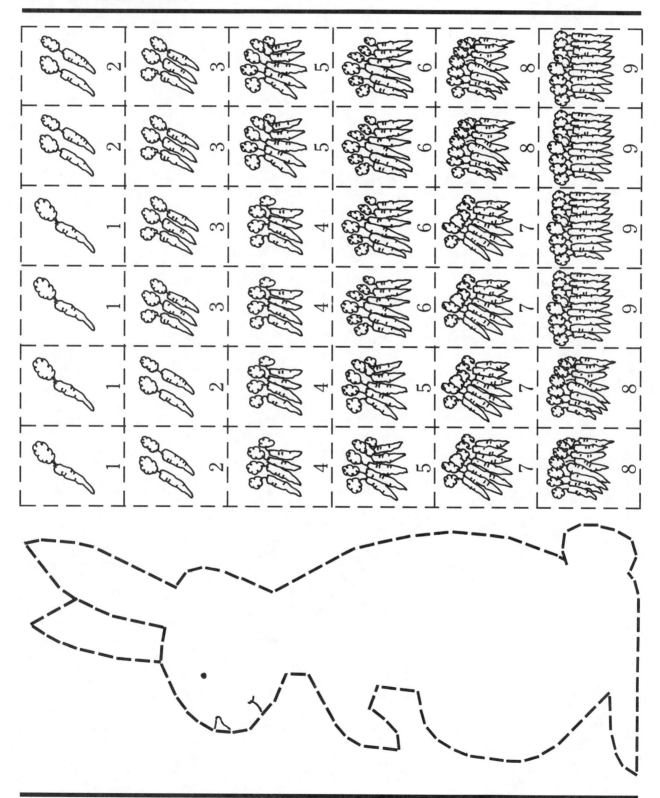

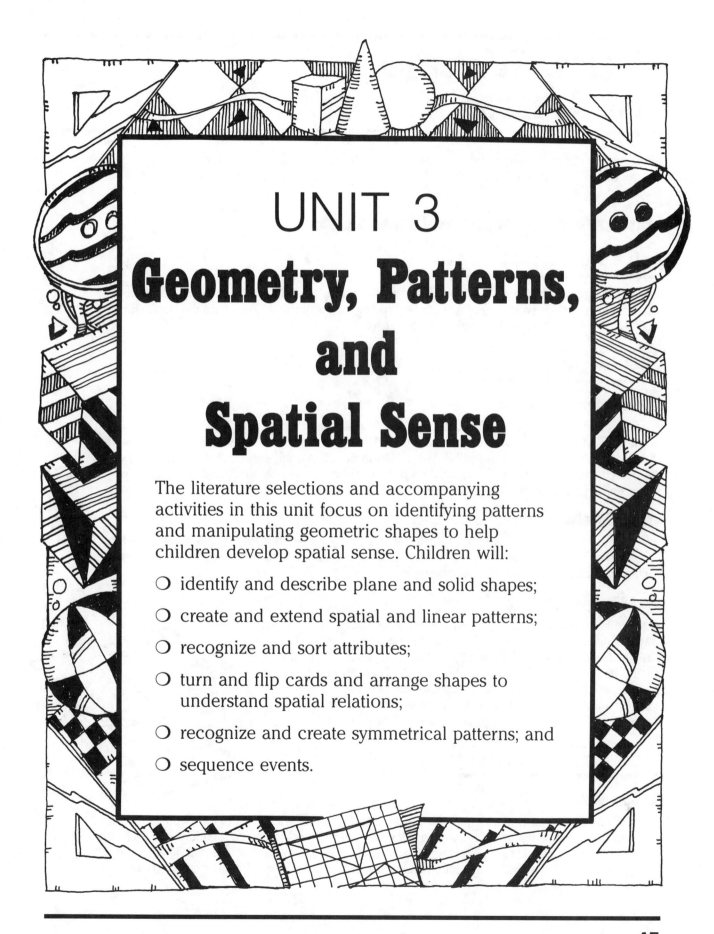

UNIT 3
Geometry, Patterns, and Spatial Sense

The literature selections and accompanying activities in this unit focus on identifying patterns and manipulating geometric shapes to help children develop spatial sense. Children will:

○ identify and describe plane and solid shapes;

○ create and extend spatial and linear patterns;

○ recognize and sort attributes;

○ turn and flip cards and arrange shapes to understand spatial relations;

○ recognize and create symmetrical patterns; and

○ sequence events.

Circles, Triangles and Squares

by Tana Hoban
Macmillan, 1974

Along city streets, in the kitchen, at the park—everywhere we look, in fact—circles, triangles, and squares are waiting to be discovered! Children will enjoy identifying these shapes in this collection of black and white photographs.

Mathematical Concept: Identifying shapes

SHARED READING

○ Ask children to close their eyes and recall their trips to school this morning. Invite children to name things they passed that resemble circles, triangles, or squares. As you share the book with children, talk about the shapes in each photograph.

MATH ACTIVITIES

Looking for Shapes

Provide children with a notepad and pencil. Take a walk around the school, school yard, or neighborhood looking for objects in which circles, triangles, or squares are visible. Ask children to sketch objects that interest them. When you return to the classroom, ask children to draw one of the objects they sketched on art paper, then paint it, outlining the triangle, circle, or square they see in dark colors. Let children discuss their work.

What Wheels!

Give pairs of children two copies of Activity Sheet 17, crayons, scissors, paste, and a large sheet of white paper. Have partners work together to design and build an amazing vehicle by coloring and pasting together circles, triangles, and squares.

Play Playgrounds

Have children work in small groups to plan a playground using straws, clay or plasticine, and recycled materials such as toilet paper tubes, berry baskets, small boxes and lids, and pieces of string. Each group will also need a large piece of corrugated cardboard or other stiff material for a base, scissors, and glue. Encourage children to incorporate circles, squares, and triangles in their playgrounds. Give each group an opportunity to present its playground design to the class.

ACROSS THE CURRICULUM

Language Arts

Place the book *Circles, Triangles and Squares* in a writing center with strips of paper. Invite children to work individually or in pairs to select several photographs, write a caption for each on a separate strip of paper, then place the captions in a shoebox labeled "Which Photograph Am I?" When everyone has written a caption, children can take turns in the learning center trying to match the captions and the photos.

Circles, Triangles and Squares

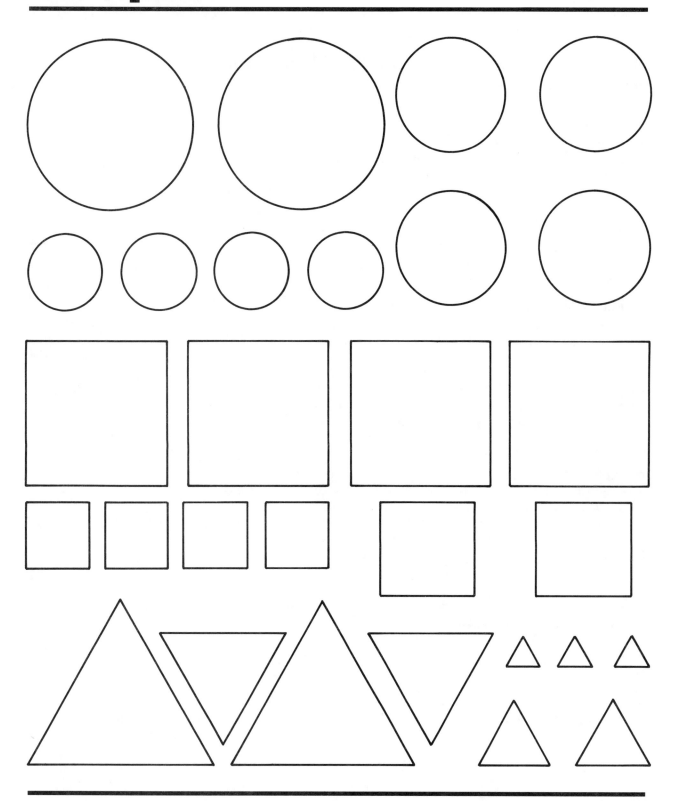

Listen to a Shape

by Marcia Brown
Franklin Watts, 1979

The language of shapes is explored in this beautiful collection of nature photographs. A mixture of captions, questions, and poetry brings the photographs to life. As children concentrate on nature's variety of forms, they will indeed begin hearing the shapes "talk."

Mathematical Concepts: Identifying and describing plane and solid shapes

SHARED READING

○ Discuss the title, *Listen to a Shape,* with children. Ask what a person might hear if he or she listened to a shape.

○ As you reread the book, encourage children to listen for words that describe a shape such as *round, curl, crescent,* and *point.* Make a list of shape-words on chart paper, and encourage children to think of other words to add to the list. Save the list for use with "Shape-Talk."

MATH ACTIVITIES

A Gallery of Shapes

Show children the pages in *Listen to a Shape* in which the photographs themselves have identifiable shapes. Give groups of four children activity sheet 18, scissors, nature magazines or seed catalogs, and paste. Let children cut out the picture frames on the activity sheet, look through the magazines to find pictures in which a circle, square, triangle, or rectangle shape can be identified, then paste these pictures on the corresponding shape frame. Decorate and display pictures on a Gallery of Shapes bulletin board.

Shape-Talk

Assemble a collection of familiar objects having these forms: sphere, cone, cube, and cylinder, and place them in a large bag. Display the sphere, cone, cube, and cylinder from a set of geometric solids. Have children describe each solid using words from the list made earlier on chart paper. Then tell children that there are some mystery objects in the bag, and each object is shaped like a solid shown. Let volunteers take turns secretly selecting an object from the bag, then giving clues to the class such as, "It is shaped like a cylinder." It can be made from glass, plastic, or paper. "What is it?" (a cup)

ACROSS THE CURRICULUM

Art

Set up a printing center of sponges soaked in tempera paint, large painting paper, and a collection of nature objects such as twigs, stones, pinecones, flowers (or seed pods), and honeycomb. Ask children to press objects on the sponges, then make prints of the objects on paper. Display dry prints. Let children look for and identify shapes in each other's nature prints.

Listen to a Shape

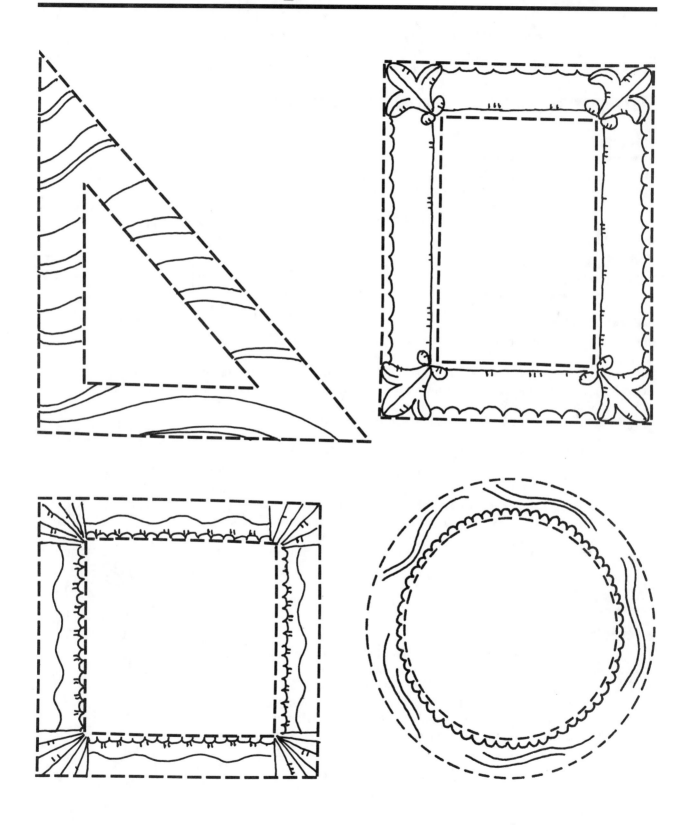

The Very Busy Spider

by Eric Carle
Scholastic, 1984

A spider is blown through a field, lands on a fence post near a farm and begins to spin a web. She works diligently through the day, despite the distractions of all the farm animals, until her work is complete. Raised illustrations let children see and feel patterns in the spider web as it forms.

Mathematical Concepts: Spatial and linear patterns

SHARED READING

○ Ask children to describe what they know about spider webs. Encourage children to talk about patterns they've seen in webs. Then read the story aloud. Let children describe the progress of the web in each succeeding illustration.
○ Give children time to look through the book and feel the raised illustrations of the spider, the web, and fly.

MATH ACTIVITIES

Animal Patterns

If children have not had prior experience with linear patterns, you might want to give them linking cubes and have them form simple patterns such as red, blue, red, blue; or red, red, blue, red, red, blue. Then give children two copies of Activity Sheet 19, scissors, crayons, paste, and strips of paper. Ask children to color, then cut out the animals. Children can form linear patterns with their animals, then paste the patterns on a big piece of paper. Let volunteers show their patterns to the class, "reading" their patterns as they display them.

Spider Web Patterns

Give each child string, scissors, paste, and a sheet of heavy paper or oaktag on which you have copied the beginning of a spider web as shown. Show children the illustrations of the web in the story and discuss the pattern. Ask children to "spin" their own webs on the heavy paper or oaktag using string and paste. Encourage children to use language such as *circle* or *circular, repeating,* and *lines* to describe the patterns they form. Save finished webs for the science activity that follows.

ACROSS THE CURRICULUM

Science/Art

Lead children in an investigation about spiders. Together, collect and display photographs and illustrations of spiders. Have children take closer looks at the spiders' legs. How do the legs help a spider crawl and climb? Give each child a lump of clay and 4 pipe cleaners that have been cut in half (8 pieces). Ask children to use their clay and pipe cleaners to make spider models. When the models are dry, children can attach them to their spider web artwork.

The Very Busy Spider

"A Lost Button" from Frog and Toad Are Friends

by Arnold Lobel
HarperCollins, 1970

After a long walk in the woods, Frog and Toad return home to find that Toad had lost a button from his jacket. The two friends retrace their steps to look for the button. Toad had lost a white, four-holed, big, round, thick button, but each button Frog and the other animal friends find does not match. Angry at Frog and his friends for not finding the button, Toad returns home to find his button on the floor. Feeling apologetic, Toad takes all the buttons that have been found in the woods and sews them all over his jacket. He gives the jacket as a gift to his good friend, Frog.

Mathematical Concepts: Recognizing attributes, sorting by attributes

SHARED READING

○ Ask children if they have ever lost a button from their clothing. Have them share descriptions of their lost buttons. Encourage children to use descriptive language in their explanations including shape, color, size, or number of holes. Then tell children that they will hear a tale of a toad who lost a button from his jacket.

○ Ask children to describe the buttons on the clothes they're wearing. Encourage children to use attributes such as shape, color, size, thickness, or number holes in their descriptions.

MATH ACTIVITIES

Sorting Rules

Make two copies of Activity Sheet 20. Have children help you color the button shapes—one sheet red and on the other sheet blue (or use any two colors). Paste the buttons on felt, then cut them out and place them on a flannelboard. Begin to sort the buttons in some way such as round and not round. Have children try to guess the sorting rule. After several rounds, make the sorting rule more difficult by using two attributes such as "round and red buttons."

Guess My Button

Use the set of buttons prepared for the activity above. Place the buttons on the flannelboard. Choose one of the buttons and draw a picture of it to record your choice. Tell children that you will be Toad and they will be Toad's friends. Let children ask questions that require yes/no answers to determine which button you have chosen. Children can use the clues to eliminate choices. For example a child might ask, "Is your button round?" After you respond, let the child come to the flannelboard and remove the buttons that do not fit the clue. Children can continue to ask questions until the correct button has been identified.

One-Difference Trains

Prepare one set of buttons for each pair of children by copying Activity Sheet 20 on two colors of construction paper. Children will also need construction paper. One child in each group places a button on the construction paper. Another child finds a button that is different in only one way, then places it next to the first button. Partners continue, in turn, to form a one-difference train. Ask children to paste their completed train on the construction paper.

"A Lost Button" from Frog and Toad Are Friends

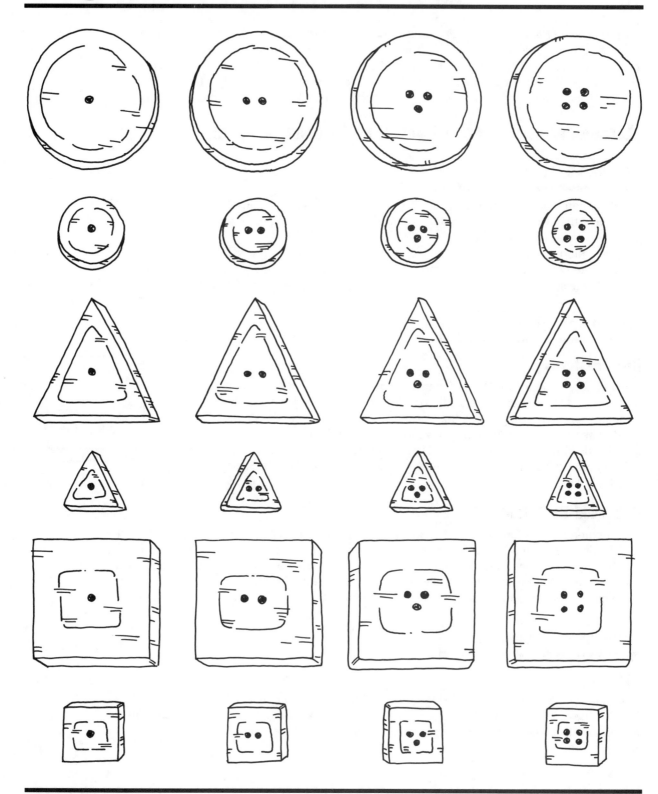

The Legend of the Indian Paintbrush

by Tomie dePaola
Putnam, 1988

In this retelling of how the Indian Paintbrush, a brilliant wildflower, got its name, an Indian boy longs to paint the colors of the sunset for his People. Every evening he goes to the top of the hill to look at the sky's colors and try and understand how to make them. A voice finally leads him back to the hill where he finds paintbrushes filled with the colors of the sunset. Little Gopher carries his painting back to his People and leaves behind brushes that take root and bloom.

Mathematical Concepts: Spatial and linear patterns

SHARED READING

○ Ask children to describe the colors they would use to paint the sunset. Then ask them to think about how they would get these colors if they could not use crayons or paints. After some discussion, read the story to children.

○ Give children paper strips cut from red, yellow, orange, and purple construction paper. Show children the colors of the sunset in the book and ask them to arrange their paper strips to resemble the sunset. Let children describe repeating color patterns in their arrangements.

MATH ACTIVITIES

Bead Patterns

Show children pictures of Native American beading. Then give each child clay and a toothpick. Let children make small round beads from the clay, then poke a hole through each bead using the toothpick. After the beads dry, children can paint the beads. Give children string with tape wrapped around the end. Ask children to string their beads in repeating patterns. On Support Master 1 (Grid Paper), children can color in the corresponding pattern. Attach beads to grid paper and display on a bulletin board.

Patterned Buckskins

Ask children to point out some of the patterns in Little Gopher's paintings. Then give each child Activity Sheet 21 and a brown paper bag (in place of the buckskin Little Gopher used). Have children cut the paper bag to make large sheets of paper, then use the patterns on the activity sheet to create the beginnings of patterns on their "buckskins." Children can use paint or markers to color and continue the patterns.

ACROSS THE CURRICULUM

Art

Ask children if they remember what the boy in *The Legend of the Indian Paintbrush* uses for paintbrushes (hairs from different animals). Let children experiment with using different items as paintbrushes. They might try leaves, sticks, pieces of paper, or blades of grass.

The Legend of the
Indian Paintbrush

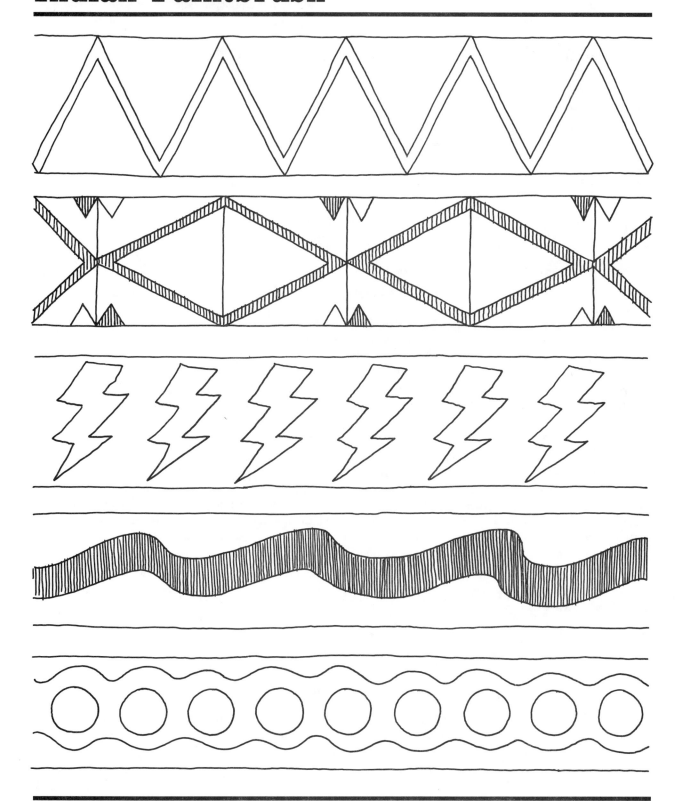

A House Is a House for Me

by Mary Ann Hoberman
Penguin, 1988

Colorful illustrations and lists in rhyme give readers an imaginative and creative look at what a house can be—from igloos for Eskimos to pods for peas.

Mathematical Concepts: Exploring shapes, sorting by attributes

SHARED READING

❍ Ask children to describe different types of houses they have seen. List children's responses on chart paper. Then read the story title and let children predict types of houses that might be included in the story.

❍ After reading the story aloud, let children continue to add "houses" to their list. Encourage children to use their imaginations.

MATH ACTIVITIES

Houses, Houses, Houses

Give small groups of children poster board (or a large sheet of paper), magazines, scissors, and paste. Ask children to look through the magazines for examples of houses. Have children cut out and sort their pictures. For example, children might group:

❍ houses for people   ❍ houses with windows
❍ houses for animals   ❍ houses with doors
❍ houses for things   ❍ houses on the water

Next have children paste the groups of houses on poster board and label each group. During sharing time, ask children to present their posters and describe their sorting methods.

Pattern Puzzles

Give children Support Master 6 (Pattern Shapes), Activity Sheet 22, crayons, and scissors. First, have children color the shapes as indicated. Ask children to choose the pattern shapes they think will fill in each house exactly. Have them experiment until they find the right combination of shapes to make each house shape. Encourage children to recognize that there may be more than one way to make each shape by having them record the color of each pattern shape they used to make the house shapes (for example, green, red, green, green, green, green, orange, orange, orange is one solution for house 2), then share their solutions with each other.

ACROSS THE CURRICULUM

Language Arts

Place the book *A House Is a House for Me* in a writing center along with paper, pencils, and crayons. On a strip of oaktag write:

A _____ is a house for _____.

Ask children to think of types of houses to complete the sentence. Children can record each sentence they create on a sheet of paper, then illustrate it. After several days, compile children's papers into a class book.

A House Is a House for Me

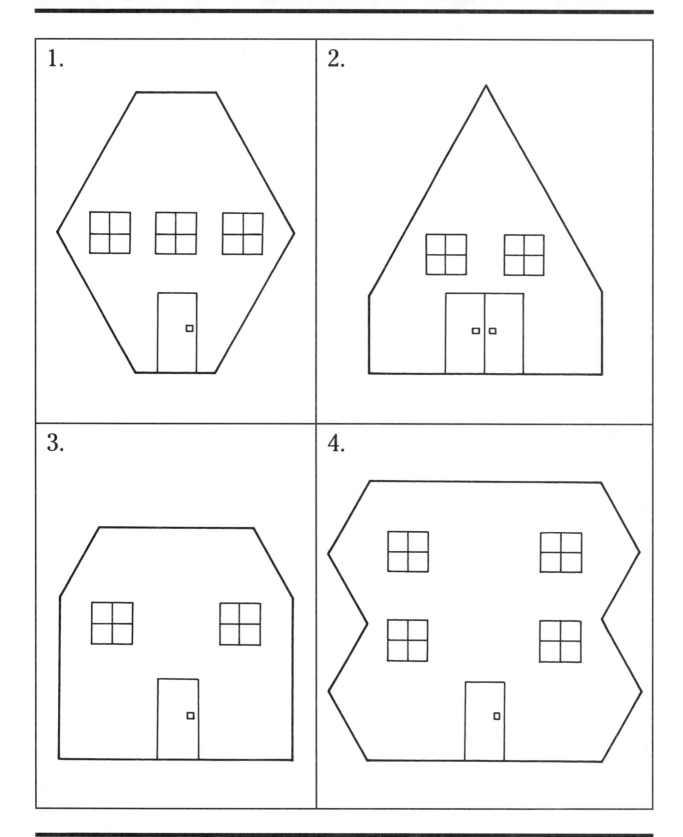

Round Trip

by Ann Jonas
Greenwillow, 1983

The technical art and vivid black and white illustrations will fascinate children. Read the story from front to back, and then magically, turn the book around and continue the story. Children won't believe their eyes!

Mathematical Concept: Understanding spatial relations

SHARED READING

❍ Discuss the concept "round trip" with children. Encourage them to think of instances when they have heard the phrase.
❍ Talk about how the views from the same car or bus window on the way to school or to a store may differ from the views on the way home.
❍ Reread the story, encouraging children to compare the "there" and "back" images. Ask children how each illustration represents day and night.

MATH ACTIVITIES

What Does It Look Like to *You?*

Give pairs of children Activity Sheet 23, scissors, and paper for recording. Ask partners to cut apart the cards, shuffle them, and place them face down in a stack on the table between them. As they turn the cards over one at a time, have each child record the card number and what the image looks like (a birthday candle, a pencil, etc.). When each card has been displayed, have partners compare their lists, turning the cards around to see each other's points of view. Challenge children to make their own drawings on blank index cards, and play the game again.

Shape Town

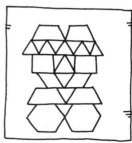

Each child needs a copy of Support Master 6 (Pattern Shapes), scissors, paste, and a sheet of black construction paper. Children cut out (uncolored) pattern shapes and arrange them to make a beautiful picture of a building on the construction paper. You may wish to show children the illustrations from *Round Trip* again to get them started. Display the buildings on a "Shape Town" bulletin board.

Make a Puzzle

Have children choose large magazine pictures, cut them out, cover the backs of them with glue, and mount them on oaktag. Help children trim away edges and demonstrate how to cut the pictures into 5 to 10 pieces to make puzzles. Have children write their names on the backs of each of their puzzle pieces and on a storage envelope. Invite kids to share their puzzles in an activity center.

ACROSS THE CURRICULUM

Language Arts/Art

Have small groups of children write and illustrate stories called "The Upside-Down Day." Designate an *Upside-Down Time* when some classroom items can be turned upside down and children can read their stories.

Round Trip

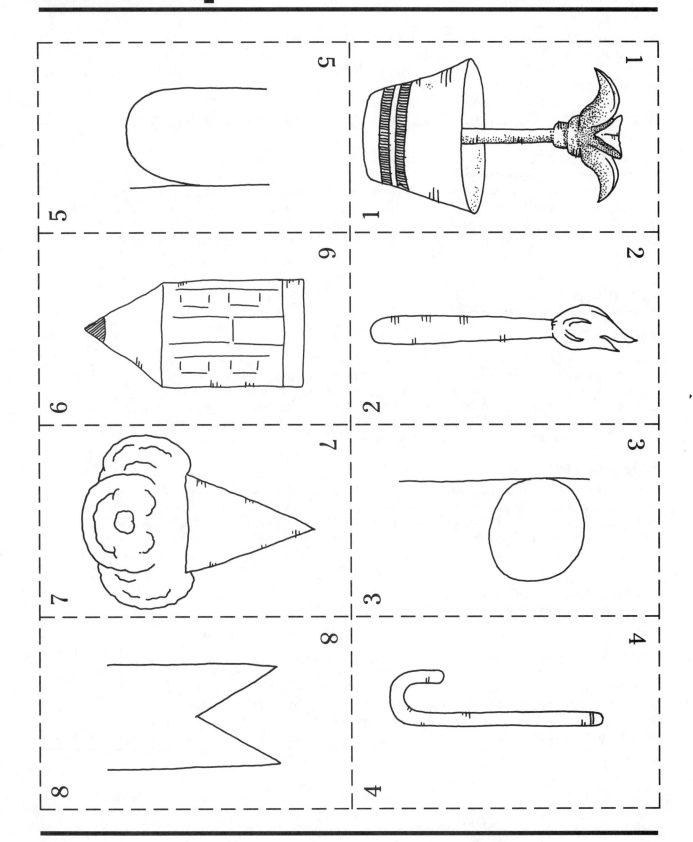

The Goat in the Rug

by Charles L. Blood and Martin Link
Macmillan, 1976

In this true story of a Navajo weaver (humorously told from the point of view of her goat, Geraldine), readers learn how mohair is shorn, washed, spun, dyed, and woven into a one-of-a-kind Navajo rug. The clear illustrations and decorative border patterns complement delightful and informative text.

Mathematical Concepts: Recognizing patterns and symmetry

SHARED READING

○ Ask children to guess what the title means. Invite children to share what they know about goats and rugs. Ask children to guess what goats have to do with rugs. Talk about how the meaning of the title would change if *on, under,* or *next to* were substituted for the word *in.*

○ After reading the story, display the pages of the book again, and have children identify all of the visual patterns they see. Ask children to identify shapes and patterns in the decorative borders.

MATH ACTIVITIES

Design a Rug

Display the illustration from *The Goat in the Rug* in which Geraldine checks the rug before Glenmae takes it off the loom. Help children identify lines of symmetry in the rug. Distribute copies of Activity Sheet 24 and Support Master 6 (Pattern Shapes) to pairs of students. Ask partners to color and cut out their pattern shapes then use them to create a rug design that demonstrates symmetry. Encourage children to experiment before finalizing their designs by pasting the pattern shapes to the activity sheet rug outline.

Body Symmetry

Ask children how their bodies are like the rug in the book. Talk about body symmetry. Ask them to identify matching body parts such as eyes, ears, nostrils, arms, legs, and so on. Seat children in pairs and ask them to work together to draw pictures of themselves, including things that are *not* symmetrical as well, such as a missing tooth, a watch on one arm, or a pocket on one side of a shirt.

ACROSS THE CURRICULUM

Art

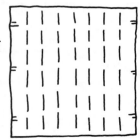

Give each child a sheet of construction paper (9 by 12 inches) and 1-by-9-inch paper strips in a variety of colors. Help them cut the construction paper as shown. Then help children weave the paper strips over and under the cuts on their construction paper, alternating the way they begin each "row" of weaving. Children should paste down the beginning and end of each row as they complete it. Laminate children's weaving to make placemats.

The Goat in the Rug

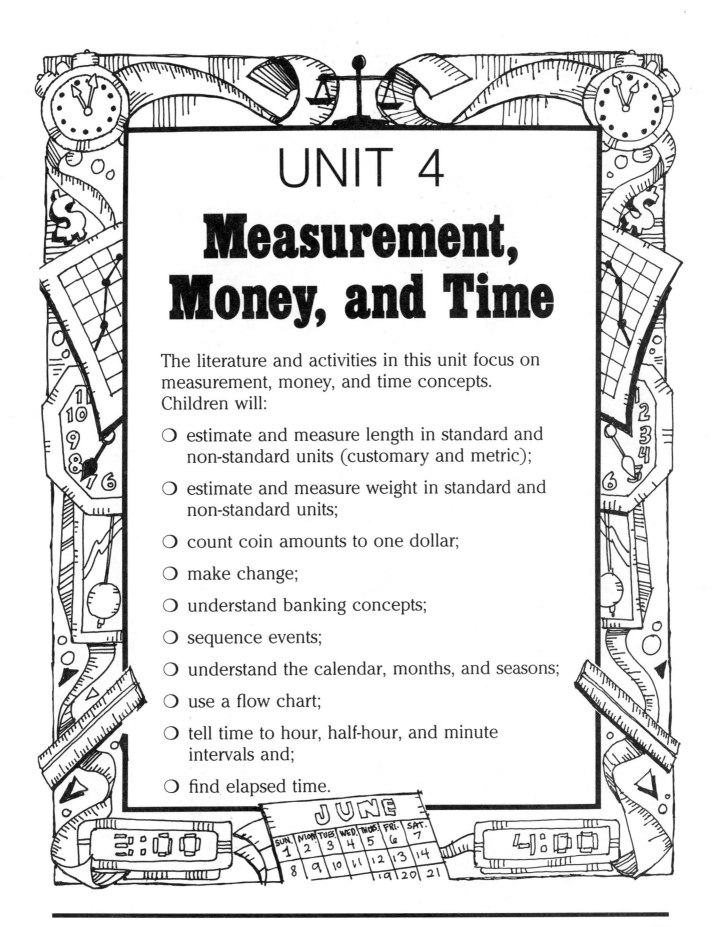

UNIT 4

Measurement, Money, and Time

The literature and activities in this unit focus on measurement, money, and time concepts. Children will:

○ estimate and measure length in standard and non-standard units (customary and metric);

○ estimate and measure weight in standard and non-standard units;

○ count coin amounts to one dollar;

○ make change;

○ understand banking concepts;

○ sequence events;

○ understand the calendar, months, and seasons;

○ use a flow chart;

○ tell time to hour, half-hour, and minute intervals and;

○ find elapsed time.

How Big Is a Foot?

by Rolf Myller
Macmillan, 1972

What birthday present can a king give to a queen who has everything? A bed, of course! (Beds hadn't been invented yet.) To find the size the bed should be, the King has the Queen lie on the floor (in her pajamas and crown) while he paces around her to find the measure. The carpenter's apprentice repeats the measuring procedure, but the bed he builds is too small for the Queen. It then occurs to the apprentice that his foot might not be the same size as the King's foot.

Mathematical Concepts: Measuring length in non-standard units, measuring length in metric units

SHARED READING

○ Ask children to listen to the story carefully to find out how using feet as a measuring tool might lead to trouble. After reading the story, let children explain how the trouble started.
○ Call on three or four volunteers to pace off the length of the classroom in "feet." Let children explain why the measures are different.

MATH ACTIVITIES

Design a Bed

Children can work in pairs to design beds for themselves. Give each child Activity Sheet 25, scissors, and crayons. Using two foot cut-outs, have one partner measure to find about how many "feet" long a bed would need to be in order for the other partner to "just fit." Children then switch roles and repeat the activity. Have children draw pictures of their beds on the activity sheets and include the "foot" measurements for the bed's length and width.

Units of Measure

Give small groups of children a pile of small paper clips and a pile of large paper clips. Ask groups to make a chain from each size clip. Children can use the two chains to make measurements in units of paper clips, such as the width of their desks or the height of a table. Talk about why it is important for everyone to use the same unit of measure.

How Long Are Our Shoes?

On a sheet of drawing paper, have each child trace around his or her shoe, cut it out and decorate it. Have children measure the length of their shoe tracings to the nearest centimeter, then exchange papers with a friend to verify the measurements. Have children record their shoe measurements on a graph.

How long are our shoes?

ACROSS THE CURRICULUM

Social Studies

Ask children to think of jobs in which workers need to make measurements of any kind. For example, postal workers weigh packages, shoe salespeople measure feet, and cooks measure ingredients. List children's responses on the chalkboard and discuss.

Name _____

How Big Is a Foot?

A Bed Fit for a King or Queen!

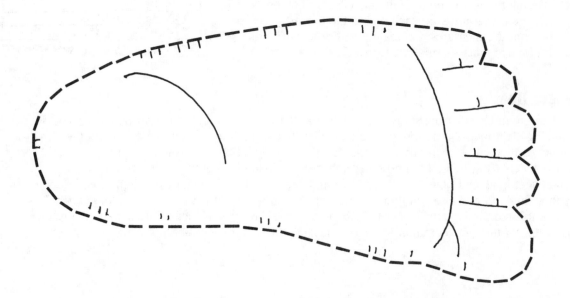

Inch by Inch

by Leo Lionni
Astor-Honor, 1962

An inchworm convinces some hungry birds that he ought not to be eaten because he is a useful measuring tool. After measuring the tails, necks, beaks, and legs of various birds, the inchworm is challenged by a nightingale to measure its song, or he will be eaten. The inch worm cleverly responds by "inching" out of the situation.

Mathematical Concept: Measuring length in standard units

SHARED READING

○ Display the cover illustration, and ask children to identify the creature pictured. Ask children to tell how they think the title of the book is related to the creature.

○ After reading the story, give each child a 1-inch piece of green yarn. Ask children to pretend their yarn pieces are inchworms. Have them place a finger at each end of the yarn, and move the yarn the way the inchworm moved. Then have them use the yarn to find the length of a book in inches.

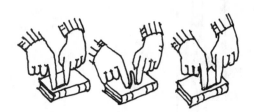

MATH ACTIVITIES

How Do Our Pencils Measure Up?

Remind children that the inchworm uses his body to measure inches. Ask them to tell what tool we use to measure inches (ruler). Provide groups of four children with an inch-ruler, and paper for recording. Have groups label two columns on the paper: *length in inches* and *number of pencils*. Ask children to gather all of the pencils that belong to the group members, measure each one to the nearest inch, and record the measures. Combine the information from each to make a class graph titled *How Do Our Pencils Measure Up?*

length in inches	number of pencils
1 inch	
2 inches	
3 inches	I
4 inches	
5 inches	II
6 inches	III
7 inches	Ⅲ I

Scavenger Hunt

Divide children into small groups. Give each group Activity Sheet 26 and a measuring tape. (A strip equaling eight inches is provided on the activity sheet. Children can cut out and mark the strip in inches.) Explain that in this scavenger hunt, children will be gathering measurements instead of objects. They'll need to find objects that are 1-, 2-, and 3-feet high, objects that are 1-, 2-, and 3-feet long, and objects that are 1-, 2-, and 3-feet around. Challenge children to look for objects they think no one else will find. Each group should record at least one object for each measurement listed on the activity sheet. At the end of the day or week, let children share their findings. Look for the most common and most unusual entries for each measurement.

Name(s) _____

Inch by Inch

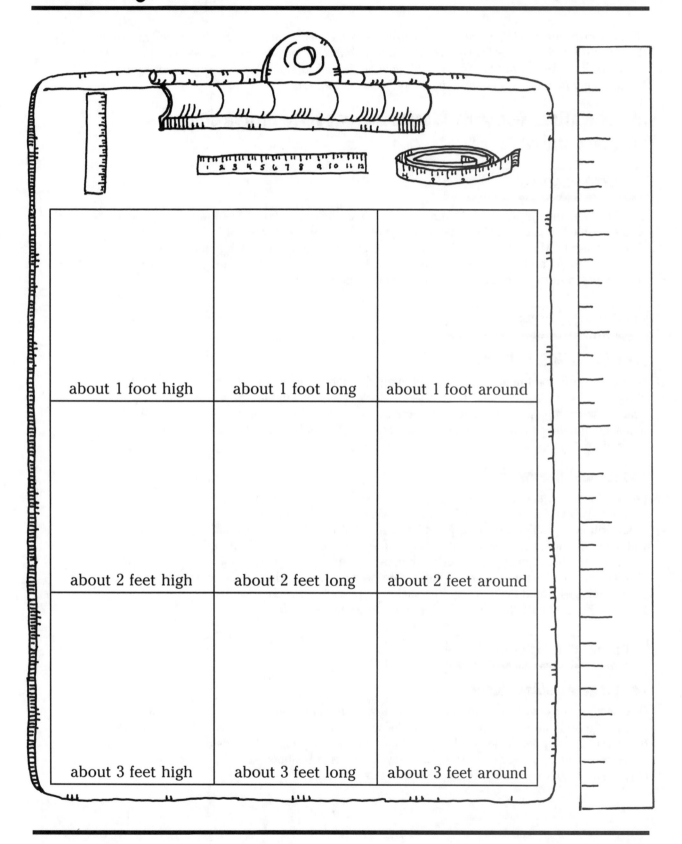

about 1 foot high	about 1 foot long	about 1 foot around
about 2 feet high	about 2 feet long	about 2 feet around
about 3 feet high	about 3 feet long	about 3 feet around

Jam
by Margaret Mahy
Little, Brown, 1986

Mr. Castle washes the dishes, cleans the house, cooks the meals, tends the garden, and takes care of the children while Mrs. Castle works as an atomic scientist. In fact, he is such a good housekeeper, that he actually runs out of things to do . . . until he decides to make jam from the plums on the Castle's plum tree. Life in the Castle household is never the same.

Mathematical Concept: Measuring capacity in non-standard and standard units

SHARED READING

○ Before reading the story, ask children to name the different flavors of jam they have tasted. As they name the jams, children may enjoy rating the flavors on a scale of 0–5.
○ Hold a Name-Mr.-Castle's-Jam Contest. Have children write original jam names on slips of paper, fold them, and place then in a plastic jar. Then compile children's suggestions in a list on chart paper, and save for use during "Shakes and Sandwiches."

MATH ACTIVITIES

How Much Will It Hold?

Give each group of three children a small margarine tub. At several work stations around the classroom, place bowls of objects such as interlocking cubes, marbles, paper clips, and wooden beads. Ask each group to measure and record how much of each material the margarine tub holds. Encourage children to record an estimate of how much each tub will hold before measuring.

Shakes and Sandwiches

Enjoy a healthy snack with your students and explore measurement at the same time. Divide the class into two groups—one to make jam sandwiches, the other to prepare frosty shakes. The jam group will need mini-bagels (or bread slices cut in half), cream cheese, and jam. The shake group will need the ingredients listed in the recipe on Activity Sheet 27.

While children prepare the snacks, ask questions about what they're doing: How many jam sandwiches could we make with all the jam in the jar? If cream cheese came in a jam jar, how many packages would the jar hold? The shake recipe calls for two cups of liquid. What's the smallest size container that we can use to make the shakes. Why?

ACROSS THE CURRICULUM

Language Arts/Art/Science

Give small groups of children magazines, scissors, paste, and a brown paper grocery bag. Ask children to find and cut out pictures of and/or words for containers, such as *tank, bathtub, drawer, bottle, bag, trunk, pool,* and *can.* Have children paste the pictures and words on the paper bag to make container collages. Store clean, dry, recycled class containers (such as yogurt cups, milk cartons, and egg cartons) in the bags for future use.

Fruit Shake

You will need:

2 cups cold orange juice
(or pineapple or grape juice)

1/2 cup powdered milk

1 drop vanilla

1 cup ice

Directions:

1. Measure ingredients.
Ask your teacher for help with the next two steps.
2. Pour all ingredients into a blender.
3. Cover and blend until ice is crushed and ingredients are mixed.
4. Pour into cups and enjoy!

(To make a fruit shake without a blender, use crushed ice. Pour all ingredients into a 1-quart container; cover and shake until mixed.)

The Turnip: An Old Russian Folktale

Rewritten by Katherine Milhous and
Alice Dalgiesh, Putnam, 1990

It was time to harvest the turnip. A procession of colorful characters pull on the turnip and on each other without success. Then the smallest of all, a little field mouse, comes to help. This timeless story reminds readers of all ages that even the smallest among us is very valuable.

Mathematical Concept: Measuring weight in non-standard and standard units

SHARED READING

○ Ask children if they have ever seen or tasted a turnip. Explain that in this story, a farmer plants some turnip seeds with interesting results. Then read the story aloud.

○ Ask children to think about why the turnip was so difficult to pull from the ground. Then display two objects with obvious weight differences such as a dictionary and a pencil. Let several children hold both objects and describe them in terms of heavy and light.

MATH ACTIVITIES

Exploring Non-Standard Units

Invite children to think of ways they could determine the weight of the turnip. Lead a discussion about scales and where children have seen them used (market, doctor's office, post office). Then let groups of children work together to determine the weight of objects in non-standard units. Give each group a balance scale (or see Activity Sheet 28 to make a simple pan balance), and objects such as interlocking cubes that can be used to balance the scale. Let children find an assortment of objects, estimate, then measure their weight in cubes, and record their findings on Activity Sheet 28.

Exploring Standard Units

Children can continue to work in groups with their pan balances. Each group will need a 1-pound weight (or an object that weighs about 1 pound). Let each child take a turn holding the 1-pound weight to become familiar with how it feels. Then ask each group to find and record other objects that weigh about one pound.

ACROSS THE CURRICULUM

Science/Art

Remind children that the turnip had to be pulled from the ground in the story. Then display a turnip or a picture of a turnip. Tell children that the edible part of this vegetable grows underground. Ask children to name the plant part as you point to the bulb (root). Let children use the classroom or school library to find other vegetables that grow as roots (for example, radishes, carrots). Have each child draw and label a picture of his or her favorite root vegetable.

The Turnip: An Old Russian Folktale

To make a pan balance, you need:
- ❍ a hanger
- ❍ 2 fruit baskets or milk cartons with tops cut off
- ❍ 4 pieces of string of equal length
- ❍ pencil or dowel stick
- ❍ masking tape
- ❍ clay or plasticine

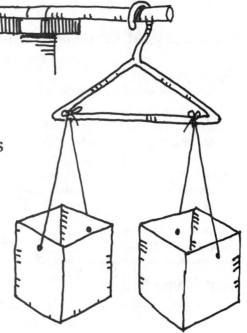

1. Poke a hole through two opposite sides of each carton. Put a length of string through each side and attach to the hanger as shown.
2. Tape a dowel stick or pencil so that about 4 inches of it is off the edge of a table or desk. Place the hanger on the stick.
3. Attach clay or plasticine inside the baskets if necessary to make them balance.

Object to be measured	Non-standard unit used to measure	Estimate	Measure

Alexander, who Used to Be Rich Last Sunday

by Judith Viorst
Macmillan, 1978

Alexander had no money—only bus tokens. Then on Sunday, Alexander's grandparents give him and each of his brothers a dollar bill. Alexander decides to save his money for a walkie-talkie. But during the week, his will to save is weakened by bubble gum, some bets, a snake rental, a garage sale, and a few other things. Now, once again, all Alexander has left is bus tokens. Alexander's humorous calamities may encourage children to be careful in spending their money.

Mathematical Concepts: Counting coin amounts to one dollar

SHARED READING

❍ Read the title and show the cover illustration to children. Ask them to imagine why Alexander might be holding his pockets inside out. Then read children the story.

❍ Give children the following play coins or coins cut from Support Master 5 (Bills and Coins)—seven dimes, four nickels, and ten pennies. Count the coins together and establish the amount as 100 cents or one dollar. Then reread the story, asking children to remove the corresponding coins as Alexander spends or loses them. Let children show how much money Alexander has left at the end of the story (none).

MATH ACTIVITIES

What Makes $1.00?

Remind children that Alexander spent his whole dollar by spending or losing several coins at a time. Tell children that there are many ways to use different combinations of coins to make one dollar. Provide groups of children with several copies of Activity Sheet 29 and Support Master 5 (Bills and Coins), scissors, paste, and a large sheet of paper. Ask group members to cut apart the Activity Sheet along the dashed lines, then work together to find coin combinations that each total one dollar. To keep the number of possibilities reasonable, limit the number of pennies that can be used in any combination to five. Children can record each combination they find by pasting the coins on the Activity Sheet charts. After children have had enough time to explore, compile their findings on chart paper as shown. Place the materials in a math center to let children continue finding combinations.

Quarters	Dimes	Nickels	Pennies
Q Q	D D	N N N N N	P P P P P

Quarters	Dimes	Nickels	Pennies
2	2	5	5
3	2	1	0
4			

$1.00 or Less

Bring in sales circulars from supermarkets or variety stores that have items priced for less than one dollar. Give groups of children a sales circular, a calculator, scissors, paste, and large sheets of construction paper. Challenge children to find as many items as they can that together total one dollar or less. Group members can cut and paste their purchases on construction paper, label each purchase with its cost, then write the total amount of the purchase.

Alexander, who Used to Be Rich Last Sunday

quarters	dimes	nickels	pennies

quarters	dimes	nickels	pennies

Arthur's Funny Money

by Lillian Hoban
HarperCollins, 1981

Violet has number problems and her brother Arthur needs money to buy a tee-shirt and cap for his frisbee team. Together they solve their problems by organizing a "bike wash." Their delightful adventure provides mathematical experiences for everyone.

Mathematical Content: Planning a fund raising event, making change

SHARED READING

❍ Show children the cover illustration of Arthur and his sister Violet. Explain that Arthur needed to earn money to buy a tee-shirt and cap. Ask children to guess what Arthur might do to earn money.

❍ Give pairs of children play money or cutouts from Support Master 5 (Bills and Coins), a calculator, and paper. Ask children to imagine they are going on a class trip. The trip will cost $10 for each group member. To raise money for the trip each group will plan a Car Wash. Write the following information on the chalkboard:

Class Trip $10 for each child

Car Wash Fund Raiser
Cars washed for $3
Supplies cost a total of $10

Ask group members to calculate the number of cars they will need to wash to raise enough money so their group can go on the trip. When children have finished, each group can share its information with the class.

MATCH ACTIVITIES

Making Change

Give groups of children Activity Sheet 30, scissors, and Support Master 5 (Bills and Coins). Ask children to cut apart the cards on the activity sheet, then fill in a price of 99¢ or less for each item. Ask each group member to select an item to buy, pretend they will pay for it with one dollar bill, then use the play money to show how much change they will receive. Group members can check each other's work. Let children paste their items and coins on construction paper.

ACROSS THE CURRICULUM

Social Studies

Ask children to recall what happened when Arthur went back in the afternoon to buy another box of soap powder. (The price went up.) Explain that prices usually do not go up in one day, but over a period of time such as several months or years. Introduce the term *inflation*. Draw the illustration shown here on the chalkboard. Let children use coins to find out how much more it will cost to buy a small box of soap powder in 1992 than it cost in 1981.

Arthur's Funny Money

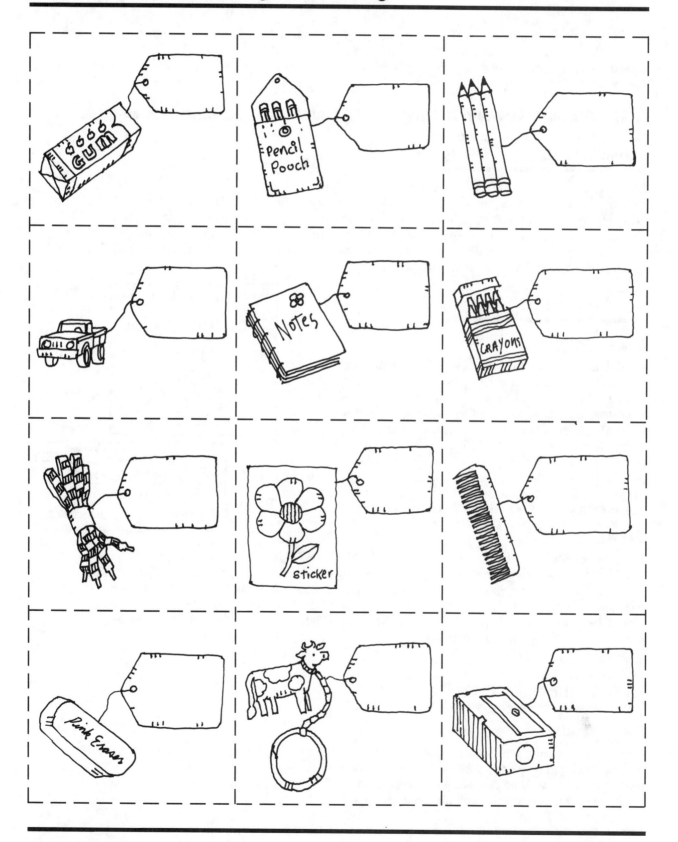

Activity Sheet 30

75

If You Made a Million

by David Schwartz
Scholastic, 1989

Explore the world of money with Marvelosissimo the Mathematical Magician as your guide. Whimsical illustrations and photographs of coins and bills will help children conceptualize equivalencies and the enormity of large numbers. The story line also provides insights into the world of banking.

Mathematical Concepts: Large money amounts, banking concepts

SHARED READING

○ Ask children to predict how high a stack of one thousand pennies would be, or a stack of ten thousand pennies. How about a stack of a million one-dollar bills? Tell children that these and other questions will be answered in the story, *If You Made a Million*. Then read the story aloud.
○ Give pairs of children play coins or Support Master 5 (Bills and Coins). Reread the first few pages of the story, and let children show the coin equivalencies for one nickel, one dime, one quarter, and one dollar as shown in the story.

MATH ACTIVITIES

Check Writing

Read aloud the section in the story dealing with checking. Give children Activity Sheet 31, scissors, and a sales circular from a local business. Children cut apart the activity sheet as indicated, then staple together the checks and the check record. Draw a large check on chart paper, and show children how to complete the check for $10.50. Then, point out the check register, showing children they have a starting balance of $20. Ask them to look through the sales circulars, find two items they would like to buy, and write separate checks for each, keeping track of the account balance on the check register.

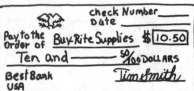

Bartering

Read the section about bartering titled "What Would We Do Without Money?" Engage children in a discussion about trading and fair trades. Then arrange children in groups of six. Give each group 12 index cards, paste, scissors, and magazines or sales circulars. Ask children to cut and paste food items on six cards and clothing items on six cards. Have children shuffle the cards and place them face down. Each group member draws two cards. Children can barter (fair trades) within their group so that each member finishes with one food item and one clothing item. Bring students together to talk about this experience with the bartering process.

ACROSS THE CURRICULUM

Social Studies

Let children research some terms associated with banking such as *deposit, withdraw, account, borrow, lend,* and *interest*. If possible, arrange for a trip to the bank, or for a banking representative to visit the classroom. If possible, obtain deposit and withdrawal slips from a local bank. Have children practice completing these slips.

If You Made a Million

Check Register					
Check number	Date	To		Amount	Balance $20.00

Check Number _____

Date _____

Pay to the Order of _____ $ [ ]

_____ DOLLARS

Best Bank USA

_____

Check Number _____

Date _____

Pay to the Order of _____ $ [ ]

_____ DOLLARS

Best Bank USA

_____

The Little Red Hen

pictures by Lucinda McQueen
Scholastic, 1985

In this classic folktale, the Little Red Hen works alone to plant, care for, thresh, and grind wheat into flour, then uses the flour to bake bread—alone and without help from her friends. Of course, when it's time to eat the bread, her friends are close at hand.

Mathematical Concepts: Sequencing events

SHARED READING

❍ After reading the story, ask children why the hen did not share the bread with her friends. Let children recount similar situations in their lives when they did all the work and felt unwilling to share the outcomes.

❍ From memory, ask children to recount the activities the hen had to perform in order to have her delicious bread. For example, finding the wheat grains, planting the wheat, threshing the wheat, baking the bread, and so on. Record each activity on the chalkboard. Then let children suggest the sequence in which the hen did these activities to produce the bread. Rewrite the activities in order on the chalkboard. (Leave this list on the board for the next activity.)

MATH ACTIVITIES

What Happened First?

Give children Activity Sheet 32, scissors, a sheet of construction paper, paste, and crayons. Ask children to cut out each picture from the activity sheet and sequence the events according to the way they occurred in the story. (If children have trouble remembering the sequence of events, they can refer to the list written earlier on the chalkboard.) Have children cut the construction paper into two strips as shown, paste them together to make one long strip, then paste the cards on the paper in sequence. Children can color the pictures, then display their sequence stories on a bulletin board.

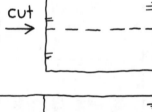

Order the Events

Write the following sequence on the chalkboard:

Picking oranges from a tree. ⇨ Cutting open the oranges. ⇨ Squeezing juice in a glass. ⇨ Drinking orange juice.

Read the sequence of events aloud with children. Let children determine if the events are in the correct sequences. Then give pairs of children index cards and crayons. Ask children to think of a sequence of events, draw a picture for each step in the sequence on an index card, and write a sentence below the picture describing the scene. Let children trade cards and sequence each other's events.

Short Time/Long Time

Draw a table on chart paper as shown. Read the headings with children and suggest an example of each, such as brushing teeth (short time), playing soccer (long time). Give each child a sticky note to record an activity from the day. Have children take turns reading their activities aloud and taping them to the chart in the corresponding column.

The Little Red Hen

She watered the small plants.

She wheeled it to the mill.

She ate the bread. Yum!

The Little Red Hen planted the grains of wheat.

She cut and threshed the large wheat plants.

She used the flour to bake bread.

Why Mosquitoes Buzz in People's Ears

retold by Verna Aardema
Penguin (Dial), 1975

In this African folktale, a mosquito tells an exaggerated story to an iguana. As the story unravels, all the jungle animals are drawn into a chain of events that results in the sun failing to rise. The tattle-tale quality of this story provides humor and a strong moral.

Mathematical Concepts: Sequencing events

SHARED READING

○ Tell children that a folktale is a story or myth that has been orally passed down to each new generation. Some folktales that children may be familiar with are *The Tortoise and the Hare, Casey Jones and the Cannonball Express,* or *King Arthur and the Knights of the Round Table.* Show children the illustration on the cover and let them guess where this folktale takes place (jungles of Africa). Then read the story aloud to children.

○ Tell children that this story has a moral—it teaches something. Invite children to explain the moral of the story in their own words.

MATH ACTIVITIES

Animal Sequence Stories

Organize students into cooperative groups. Give each group one copy of Activity Sheet 33. Instruct groups to cut out the animal cards, shuffle them, and place them face down on a table or desk. Next, ask a member of each group to turn over the cards, one at a time, as a second member records the emerging sequence in the form on a list; for example: *1. fly; 2. bluebird; 3. cat; 4. dog; etc.* After all the cards have been turned over, challenge students to work in their groups to write an original story (modeled on *Why Mosquitoes Buzz in People's Ears*) in which each of the animal characters is introduced in the same order as their list. After all the tales are complete, invite the groups to share their "sequence stories" with the class. How are they alike? How are they different?

Time Line

Invite children to name various activities they participate in during the day (reading, math, language arts, science, recess, gym, and so on). As children name activities, list them on chart paper or on the chalkboard. When a comprehensive list is recorded, arrange children in four groups and give each group a length of string, and one paper square and paper clip for each activity on the list. Ask each group to work cooperatively to draw a picture and write the name for each activity on a paper square. Group members can establish the sequence of the day's activities, then display the pictures in sequence by attaching them to the string with the paper clips. Display the four time lines and discuss any discrepancies in sequencing.

Why Mosquitoes Buzz In People's Ears

fly	cat	bee
dog	squirrel	bluebird
chipmunk	cricket	robin

Chicken Soup With Rice

by Maurice Sendak
HarperCollins, 1962

This wonderful, nonsensical rhyme for each month of the year reflects the changing seasons along with characteristics and celebrations of each month. Young readers will enjoy the rhyme, rhythm, and repetition in each stanza of the poem.

Mathematical Concepts: Understanding the calendar and months, creating a flow chart

SHARED READING

❍ Recite the months of the year aloud in unison. Then, as you read the poem, ask children to anticipate which month comes next. Ask children to identify details in the poem that describe each month.

❍ Reread the poem, letting children look more closely at the illustrations. Discuss each illustration and how it represents a stanza of the poem.

MATH ACTIVITIES

Charting the Months

Let children experiment with flow charts to arrange months in order. Distribute Activity Sheet 34, scissors, and paste. Discuss sequencing, pointing out how to follow the arrows to read the chart. Children can begin with January or any month. After children complete the chart, let pairs of children check each other's work.

How Many Days?

Show children the knuckle method for finding the number of days in each month. Children make two fists as shown— knuckles represent months with 31 days. Spaces between knuckles represent months with 30 days and February with 28 days. Working in pairs, one child can make two fists while the other recites the months in sequence, pointing to the knuckles to help tell the number of days in each month.

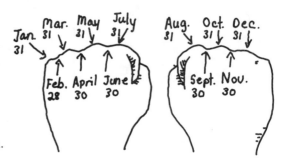

Cooperative Calendars

Let children learn about calendars by looking at old calendars and creating new ones. Tear the pages of an old calendar apart. Divide children into 12 groups and have each pick one of the old calendar pages. Have each group create a new calendar page for that month. Provide current calendars to assist students in correctly numbering the calendar squares.

Have students note special school days, holidays, birthdays, and so on. On the reverse of each calendar page, have students create art for the following month's calendar page (When the calendar is hung, the reverse of one calendar page becomes the art for the next page.) The group doing January will do two pieces of art—one for January and one for February. Put the pages together to make a new class calendar.

Chicken Soup With Rice

1. Cut out the months.
2. Arrange them in order on the flow chart.
3. Paste in place.
4. Explain your order to a friend.

March	October	December	February
November	January	July	June
May	August	April	September

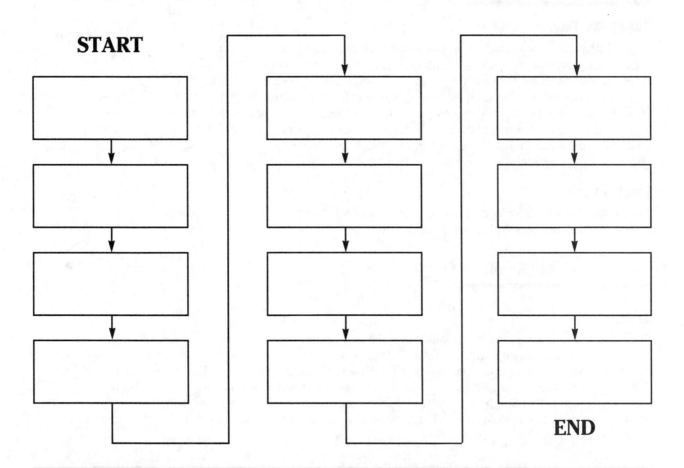

START

END

The Sun's Day

by Mordicai Gerstein
HarperCollins, 1989

In this enchanting story, a brilliantly colored sun sweeps across the sky. With each passing hour, the sun takes the form of such daily routines as a morning slice of toast, a furry orange cat taking an afternoon nap, and a pot of simmering soup ready for supper. Beautiful prose helps children associate a time of day with each daily event.

Mathematical Concepts: Telling time to the hour, sequencing events

SHARED READING

○ Name a time of day, such as 8 o'clock in the morning or 6 o'clock in the evening and ask children to describe what they do at these times. After some discussion, show children the cover illustration and ask them to imagine what the sun does during the day. Then tell children that this author imagined that the sun takes a part in the activities it sees. Ask children to look carefully for the sun in each illustration and think about what it is doing.

○ Give children small clocks with moveable hands. Reread the story and ask then to show each time on their clocks.

MATH ACTIVITIES

About My Day

Invite children to make books about their day. Give each child two copies of Activity Sheet 35, scissors, and crayons. Children are to cut apart each sheet so they have eight pages total. Set an alarm clock to ring every hour. When the alarm rings, (beginning at 9 A.M.) children stop what they are doing and record on one of their pages the time and what they are doing. Stop at the end of the school day, or give children additional pages to take home for after-school hours. Let children sequence their pages, and compile them into a book, then add a cover.

Clock Puzzle

Draw a large clock on a piece of oaktag. Cut the clock into 8 to 10 puzzle pieces. Let children reassemble the clock puzzle.

ACROSS THE CURRICULUM

Science

Talk about the sunrise and sunset. Explain that the day with the greatest amount of sunlight is on or near June 21 and the day with the least amount of sunlight is on or near December 21st. Check the weather section of a newspaper and find the time of the sunrise and sunset for a particular day. You might want to round these times to the nearest five-minutes or half-hour. Give children paper and crayons and ask them to draw a clock to represent each time (sunrise, sunset). Have children use their clocks to figure out the number of hours of daylight and the number of hours of darkness.

The Sun's Day

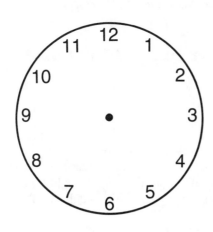

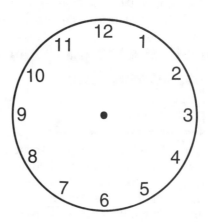

_____

_____

_____

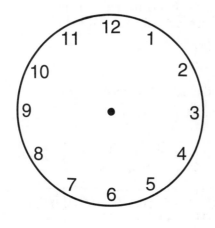

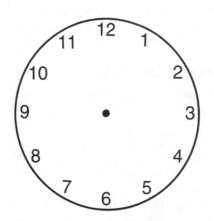

_____

_____

_____

Clocks and More Clocks

by Pat Hutchins
Macmillan, 1970

Mr. Higgins has clocks all over his house. But he is puzzled as to why the clocks do not show exactly the same time when he moves from room to room. With the assistance of the clockmaker and his watch, Mr. Higgins learns about elapsed time.

Mathematical Concepts: Time to the hour, half-hour, and minute intervals; elapsed time

SHARED READING

○ Ask children to tell where the clocks are in their houses. Let children explain how clocks that are set at exactly the same time could differ as they go from room to room to read them. Then show children the house pictured on the cover of the book and talk about the time it might take to go up and down the stairs. After reading the story, ask children to explain if Mr. Higgins really needed a watch to tell if all his clocks were telling the right time.

○ Display a demonstration clock, then read the story again. Volunteers can set the clock to show each time mentioned. Discuss the passage of time that occurs as Mr. Higgins moves from room to room.

MATH ACTIVITIES

One-Minute Books

Let children work with a partner. Give each child a copy of Activity Sheet 36 and scissors. Children estimate how many times they can perform the given task in one minute, then time the activity to check their guesses. Children can color in the pictures of people performing each activity, then cut and staple the pages together to make books. Let children add to the book during the year by timing other events.

One Hour Earlier or Later

Groups of children will need Support Master 2 (Gameboard), a die, 28 index cards or paper squares, and a game marker. Next, help children make cards as follows.

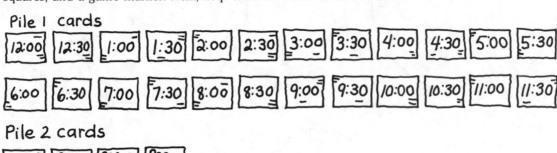

Pile 1 cards

| 12:00 | 12:30 | 1:00 | 1:30 | 2:00 | 2:30 | 3:00 | 3:30 | 4:00 | 4:30 | 5:00 | 5:30 |
| 6:00 | 6:30 | 7:00 | 7:30 | 8:00 | 8:30 | 9:00 | 9:30 | 10:00 | 10:30 | 11:00 | 11:30 |

Pile 2 cards

| one hour earlier | one hour earlier | one hour later | One hour later |

Place cards in two piles. To play, children take turns picking a card from each pile and naming the time one hour earlier or one hour later as directed. If group members judge the response to be correct, the player rolls the die, then moves his or her marker on the gameboard. Players return cards to the bottom of each pile, face down. The winner is the first to reach finish.

Clocks and More Clocks

One Minute Book

This book shows how many times I can do each activity in one minute.

By _____

Hop on one foot.

Guess: _____ times
Timed: _____ times

Say the alphabet.

ABCDEFG

Guess: _____ times

Timed: _____ times

Clap my hands.

Guess: _____ times

Timed: _____ times

Count to 20.

1 2 3 4 5

Guess: _____ times

Timed: _____ times

Touch my toes.

Guess: _____ times

Timed: _____ times

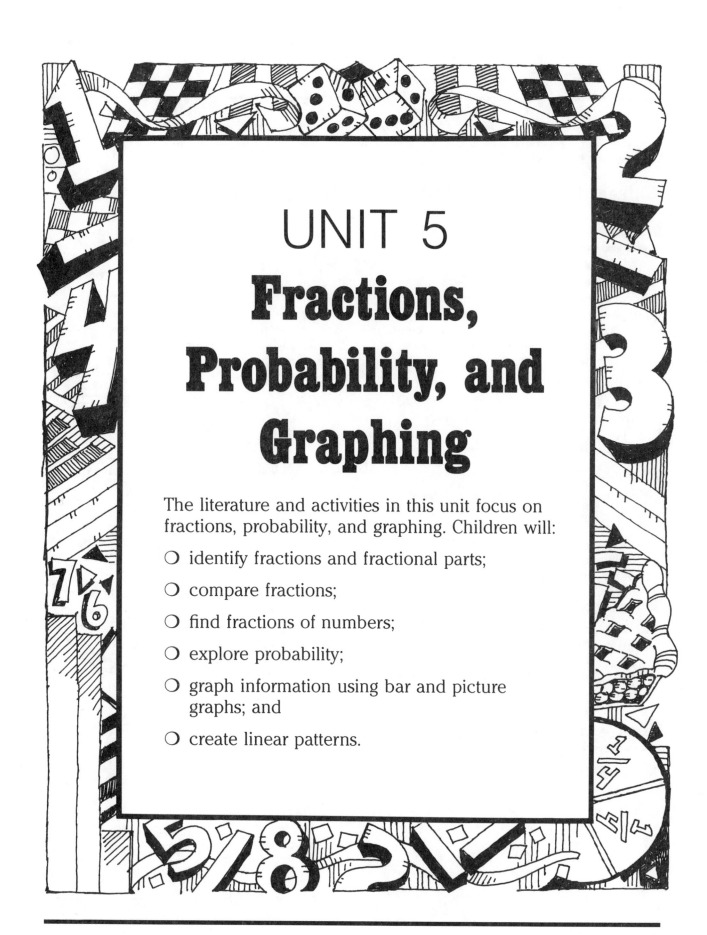

UNIT 5
Fractions, Probability, and Graphing

The literature and activities in this unit focus on fractions, probability, and graphing. Children will:

- ❍ identify fractions and fractional parts;
- ❍ compare fractions;
- ❍ find fractions of numbers;
- ❍ explore probability;
- ❍ graph information using bar and picture graphs; and
- ❍ create linear patterns.

Eating Fractions

by Bruce McMillan
Scholastic, 1991

Vivid photographs of mouth-watering foods such as bananas, muffins, pizza, and strawberry pie introduce the concepts of halves, thirds, and fourths. Recipes are included so that you can prepare your own fraction feast.

Mathematical Concepts: Identifying fractions and fractional parts

SHARED READING

○ Discuss with children how they would share food such as a pizza, an apple, or cake equally among four children. In their discussion, listen for an indication that children understand the concept of equal shares. Then show children the book and discuss the concepts of wholes, halves, thirds, and fourths presented in each picture.

○ Give each child these shapes cut from construction paper: yellow rectangle (banana), brown circle (muffin), larger red circle (pizza), and a larger yellow rectangle (corn on cob).

Reread the story and let children fold their paper cutouts to match the equal shares indicated in the photographs. Let children cut apart each shape along the creases, then match them to show "equal parts."

MATH ACTIVITIES

banana   muffin   pizza   corn on the cob

Fractional Foods

Let children look through magazines or food circulars for illustrations or photographs of food that can be shared equally. Children can decide how many equal portions they want to divide the food in their pictures into, fold the pictures, then draw lines along the folds to show equal parts. Help children label the parts of each picture with a fraction. Display children's pictures on a bulletin board.

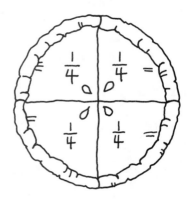

Pizza Party

Each child will need Activity Sheet 37, crayons, scissors, paste, and four half-sheets of paper. Ask children to color and cut out the food parts at the bottom of their activity sheets, cut out the pizzas, "prepare the orders" as indicated, then paste them on half-sheets of paper. Ask children to write sentences describing their pizzas, as shown.

ACROSS THE CURRICULUM

Health/Nutrition

Prepare a "fractional snack" using one of the recipes at the back of this book. Before you cut the snack into equal parts, invite children's explanations of how the snack should be divided so that everyone gets the same amount.

One half of the pizza has pepperoni.

Eating Fractions

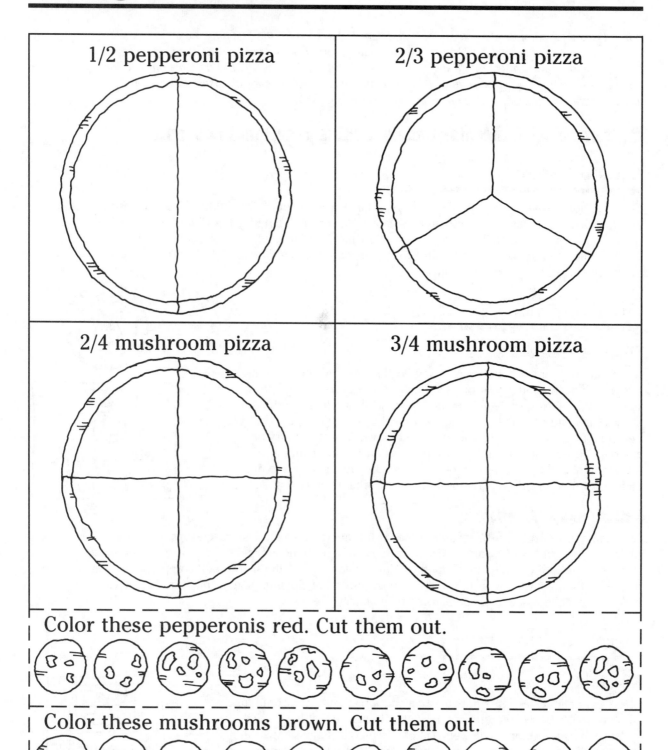

1/2 pepperoni pizza

2/3 pepperoni pizza

2/4 mushroom pizza

3/4 mushroom pizza

Color these pepperonis red. Cut them out.

Color these mushrooms brown. Cut them out.

The Half-Birthday Party

by Charlotte Pomerantz
Houghton Mifflin, 1984

When Daniel's six-month-old sister Katie stands up on her own for the first time, he is so pleased that he makes her a half-birthday party. Daniel invites Mom, Dad, Grandma, and the neighbors to the party and asks them each to bring half a present. But Daniel was so busy with his preparations, he forgets to plan his own present for Katie. The present Daniel finds for his sister that evening expresses the warmth and love in this brother-sister relationship.

Mathematical Concepts: Understanding fractional concepts

SHARED READING

❍ Ask children to tell their ages. Encourage children who include "1/2" in their descriptions to explain what that means. Tell children that Daniel plans a "half-birthday party" for his younger sister, Katie. Let children tell how old Katie is. (six months) The read the story.

❍ Ask children to think about what fractional part of a year one month is. (1/12) Then ask them to find their ages in years and months.

MATH ACTIVITIES

"Half" a Fruit Salad

Ask children to pretend they are making "half snacks" to serve at Katie's party. Give each child a pear or peach half, a strawberry or a grape, some peanut halves, some greens such as spinach or lettuce, and plastic knives. Let children use these materials to make faces. (The pear or peach half can be placed on the plate, face down. The strawberry or grape can be cut in half for the eyes, peanut halves can be arranged for a mouth, and the greens can be used for hair.) Ask children to draw pictures of their creations, write descriptions of what they used, then eat!

I used a pear half and strawberry halves and peanut halves.

Pattern Shape Fractions

Give children Activity Sheet 38, Support Master 6 (Pattern Shapes), scissors, crayons, and paste. After children color in the pattern shapes as indicated, have them explore the shapes to find those that are halves, thirds, or sixths of another. Have children paste the shapes in the appropriate places on their activity sheets. If you have an overhead, let children demonstrate for classmates how they matched pattern shape pieces to the activity sheet shapes.

ACROSS THE CURRICULUM

Science

Give each child a small brown paper bag and a pad and pencil, and take a nature walk. Ask children to look for things in nature that have equal parts, such as leaves or wildflowers. Let children collect any samples they can, and sketch pictures of anything they cannot collect. Let children paste objects between two sheets of waxed paper, then write a description as shown.

This flower has four petals. All the petals are the same.

The Half-Birthday Party

Paste 2 equal-size pattern shapes on these shapes.

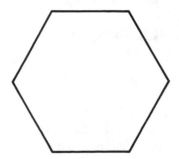 

Each shape I pasted is _____ of the whole shape.

Paste 3 equal-size pattern shapes on these shapes.

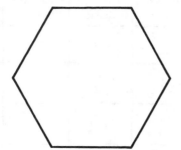 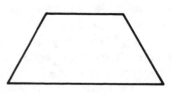

Each shape I pasted is _____ of the whole shape.

Paste 6 equal-size pattern shapes on this shape.

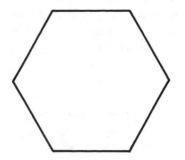

Each shape I pasted is _____ of the whole shape.

Lucy and Tom's 1. 2. 3.

by Shirley Hughes
Penguin, 1987

Lucy and Tom spend all day Saturday preparing for Granny's 60th birthday celebration. As they play at home, walk through the neighborhood, and shop for food and gifts, they continually solve problems and make important quantitative discoveries about their world. Children are sure to recognize familiar sights and situations as they accompany Lucy and Tom through a busy day.

Mathematical Concepts: Finding fractions of numbers, exploring probability

SHARED READING

○ Display the title page and ask children to guess what *1. 2. 3.* in the title might mean. Have children think of other sets of objects that come in ones, twos, or threes. Then read the story aloud.

○ Reread the story, and give children opportunities to talk about how their own experiences resemble or differ from Lucy's and Tom's.

MATH ACTIVITIES

A Fraction of the Fish

Display the page in the story in which Lucy and Tom are sharing candies. Talk about how many of the candies Lucy and Tom will each get when they share them equally (5), and what fraction of the candies each child's portion represents (1/2). Then give pairs of children 6 muffin cups or paper circles, 12 goldfish crackers, and a copy of the record sheet shown on the right. Ask children to use the muffin cups to pretend to divide the crackers among 2, 3, 4, and 6 friends, then tell how many goldfish each friend will receive and what fraction of the total number each will receive.

Number of Children in group	Number of Goldfish	Number each child will get	Fraction each child will get
2	12		
3	12		
4	12		
6	12		

Tabby or Spotted Kitten: Which Is More Likely?

Show children the page in the story on which Mopsa's five kittens are pictured. Ask them to imagine Tom reaching into the box and picking up one of the five kittens in the dark. Would he be more likely to pick up a tabby or a spotted kitten? To find out, have pairs of children place three white cubes (spotted kittens) and two brown cubes (tabby kittens) in a bag. Children can take turns removing a cube from the bag. Ask children to continue for 50 tries, then report their findings to the class.

What's the Probability of Landing on a Square?

Give pairs of children Activity Sheet 39, scissors, and tape. Introduce the figure on the activity sheet as a *cuboctahedron* and have children assemble the solid, as shown. Then ask children to determine the chances that the cuboctahedron will land on one of the square faces when rolled. Have pairs of children roll the solid 100 times and record their results with a tally. Children can report their findings in these terms: *The chance of landing on a square face is _____ out of 100.* Help children evaluate why each group achieved similar results.

Lucy and Tom's 1. 2. 3.

1. Cut along the solid lines.
2. Fold along the dashed lines.
3. Tape the tabs together to form the cuboctahedron.

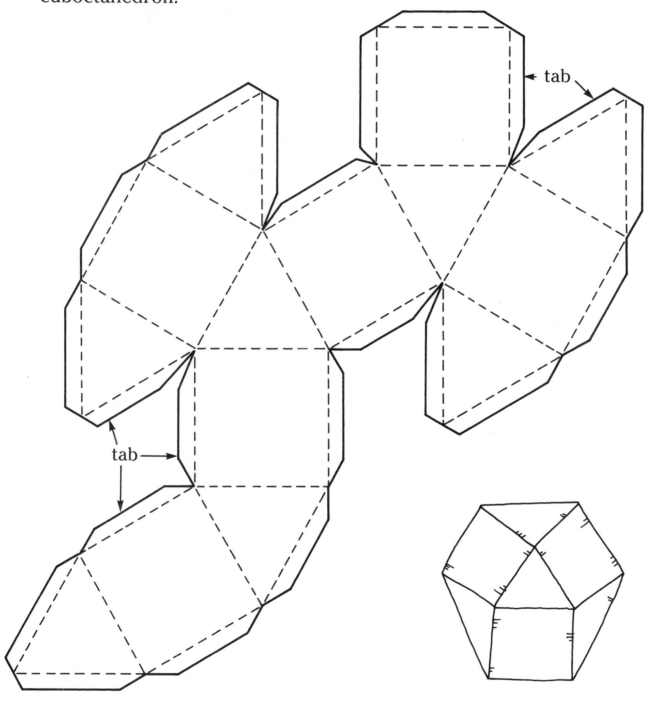

tab

tab

Caps for Sale

by Esphyr Slobodkina
HarperCollins, 1987

A cap peddler ventures into the countryside with all his wares on his head. When he unwisely leans against a tree to take nap, he wakes to find all the caps, except one, in the hands of mischievous monkeys in the tree. The peddler is unsuccessful in retrieving his caps until he throws his own cap to the ground. This wonderful story appeals to young children because of its humor and simplicity.

Mathematical Concepts: Organizing information in a graph

SHARED READING

○ Show the cover illustration and encourage children to use picture clues to explain what a *peddler* is. Before beginning the story, ask children to guess how this peddler might carry his caps as he walks around selling them.

○ After reading the story, give pairs or children color counters to represent the gray, brown, blue, red, and checked (black) caps. Ask children to count to find the total number of caps the peddler wore on his head.

MATH ACTIVITIES

Graphing the Caps

Tell the children that another cap peddler sold these caps: 6 red, 3 blue, 5 brown, and 4 yellow. Give children Activity Sheet 40, crayons, and scissors. Help them color the caps as indicated above, then color the caps below the graph as marked. Children should arrange the caps on the graph in the corresponding columns. You might want children to paste the caps in place to make a picture graph, or remove the caps, one at a time, and color to make a bar graph. Ask children questions about their completed graphs such as, *How many red caps?*(6) *How many yellow caps?* (4) *How many caps altogether?* (18) *Which color does the peddler have the most of?* (red) *Which color does the peddler have the least of?* (blue)

Patterning Caps

Let pairs of children use their caps to create linear patterns. (If children pasted their caps on the graph described above, they will need additional copies of Activity Sheet 40.) Give children Support Master 1 (Grid Paper). Ask them to create 2-, 3-, and 4-color patterns with their caps, then record each pattern on the grid paper. Discuss the patterns children create.

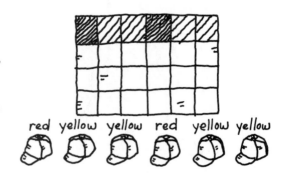

red   yellow   yellow   red   yellow   yellow

ACROSS THE CURRICULUM

Art

Give children a variety of art materials such as construction paper, paints, crayons, tissue paper, craft sticks, and yarn. Ask children to create something they would want to sell and assign prices of $5 or less. Let children use Support Master 5 (Bills and Coins) to buy and sell their work, paying for each object with the correct amount of play money.

Name _____

Caps for Sale

red

blue

brown

yellow

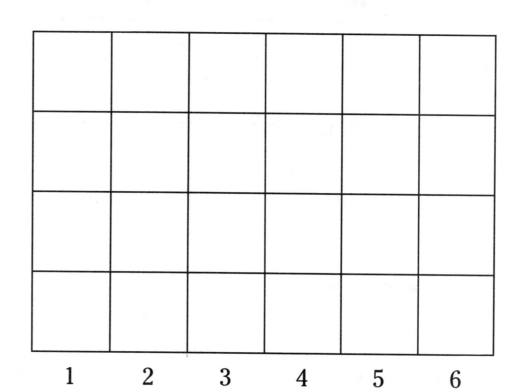

   1   2   3   4   5   6

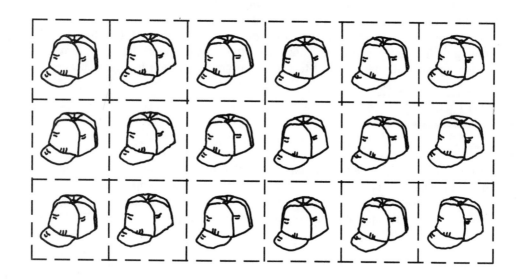

Is It Rough? Is It Smooth? Is It Shiny?

by Tana Hoban
Greenwillow, 1984

Vivid illustrations let readers visually "feel" the texture of each object—the roughness of the elephant's hide, the smoothness of the red apples. Other senses come into play as we "taste" the freshly baked bread, "smell" the aroma of the soft pretzel, and "hear" the raindrops as they hit the windowpane.

Mathematical Concepts: Constructing picture and bar graphs

SHARED READING

❍ Let children describe objects they associate with the words *rough, smooth, shiny*. List children's responses on chart paper. Add other descriptions of texture such as soft, hard, or sticky. Then, as you show the photographs in the book, ask children to describe the texture of each object.

MATH ACTIVITIES

rough	smooth	shiny
elephant	gum	pennies
beard		apples
pretzels		

Graphing Textures

Give groups of children construction paper, 2-inch paper squares, crayons, and paste. As you display the pictures again, ask children to draw and label each object on a paper square. Have children use the squares to create picture graphs, grouping the objects as rough, smooth, or shiny. Let groups display and compare graphs.

Graphing Leaves

Collect leaves from the ground. Arrange children in groups and give each a handful of leaves and Activity Sheet 41. Have children examine leaves for similarities and differences and agree on six ways to categorize them (for example, by shape, size, color, or edge patterns). Children can draw pictures representing their categories (or use category on the activity sheet), label the drawings, and color in the appropriate number of boxes to create bar graphs of their leaf groupings.

ACROSS THE CURRICULUM

Art/Science

Recycle leaves from the previous activity by making nature prints. You'll need poster paint, paintbrushes, and thin, white paper. Children can help by covering their work surface with newspaper. Have children follow along as you demonstrate the following steps:
❍ set the leaf on the newspaper and paint the entire surface of the leaf;
❍ remove the leaf gently;
❍ place the leaf paint-side up on the newspaper;
❍ place a sheet of white paper on top and rub over the leaf to make the print;
❍ remove the top paper, let dry, and display.

Is It Rough? Is It Smooth? Is It Shiny?

Types of Leaves

	1	2	3	4	5

Use any of these labels for the graph or make your own.

palm-like	feather-like	smooth-edge	serrated-edge
grass-leaf	needle-shaped	oval	pod-shaped

Support Masters

○ Master 1: Grid Paper

○ Master 2: Gameboard

○ Master 3: Place Value Models

○ Master 4: Number Cubes

○ Master 5: Coins and Bills

○ Master 6: Pattern Shapes

○ Master 7: People Counters

○ Master 8: Teddy Bear Counters

○ Master 9: Bug Counters

Grid Paper

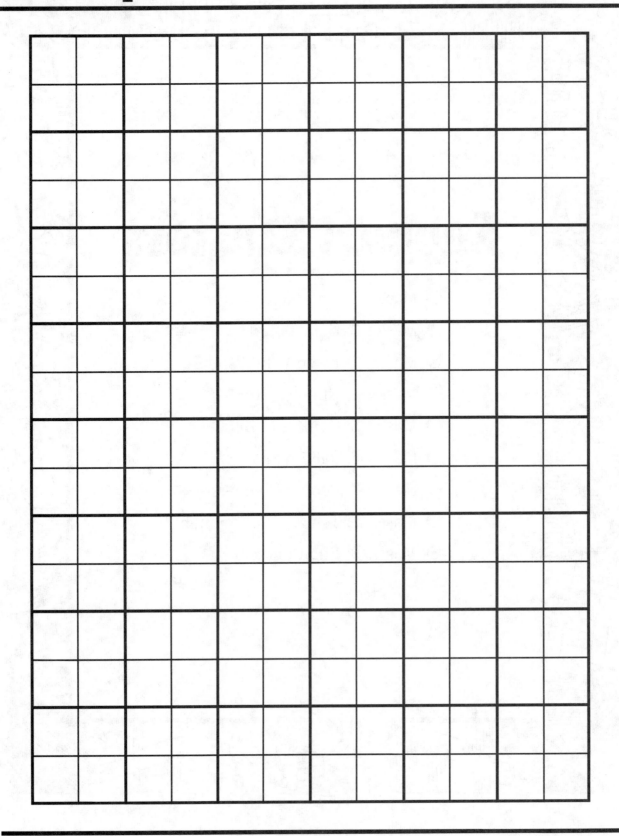

Gameboard

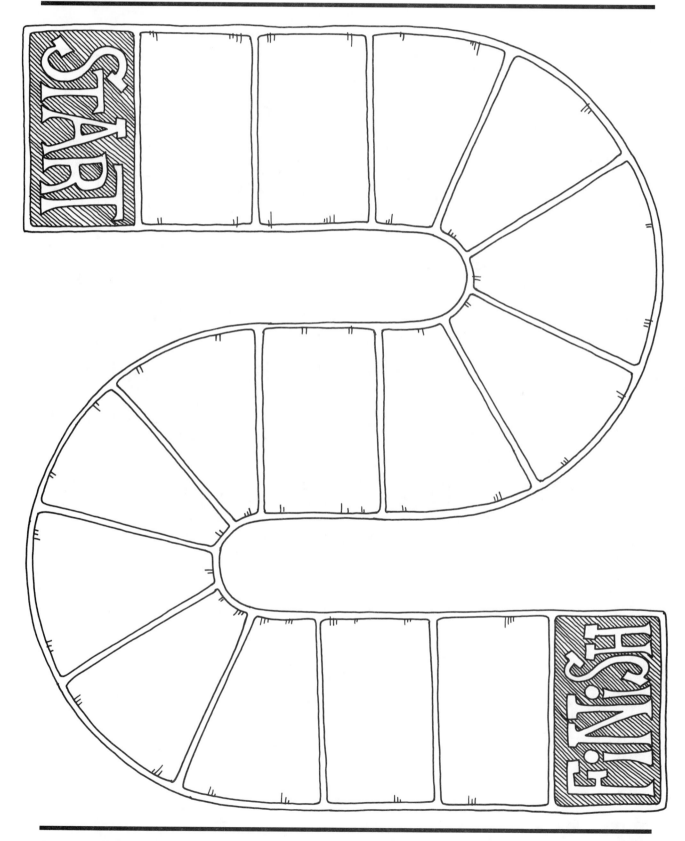

Place Value Models

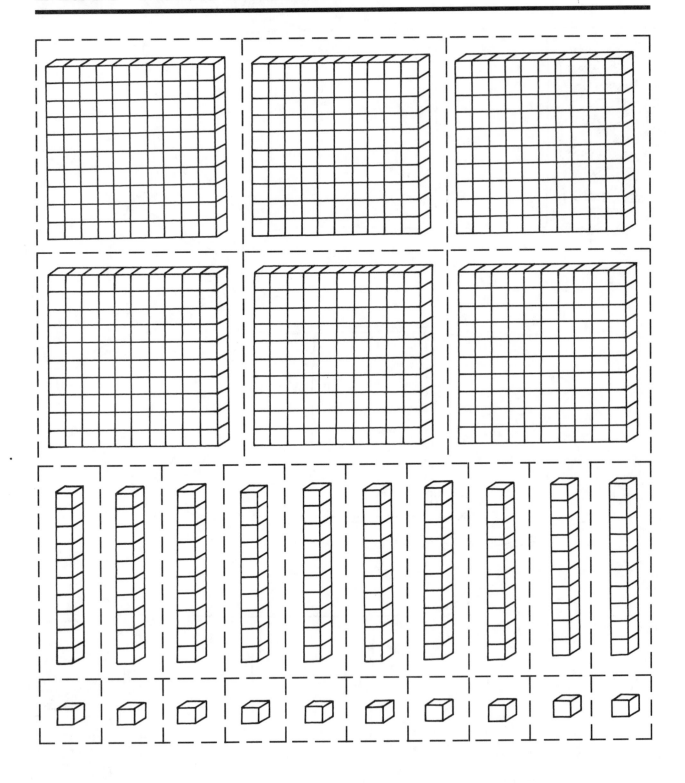

Number Cubes

1. Cut out the cubes along the solid lines.
2. Fold along the dashed lines.
3. Tape or paste the tabs to form cubes.
4. For blank cubes, reverse the folds so the numbers will be on the inside.

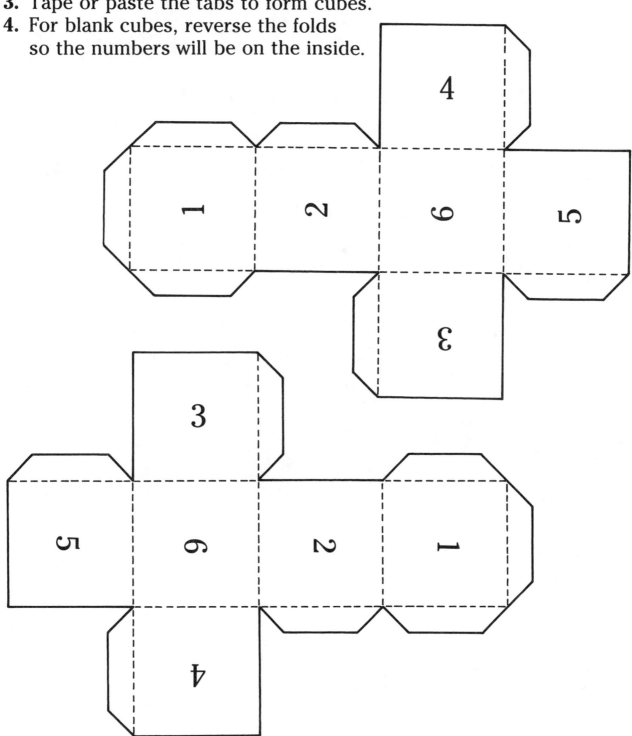

Bills and Coins

Pattern Shapes

Color yellow.

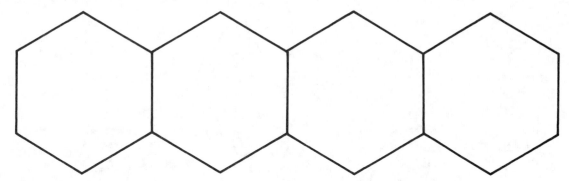

Color orange.

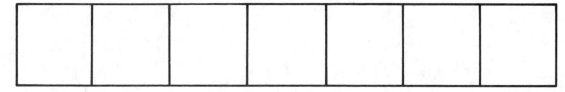

Color green.

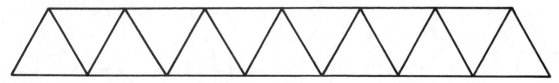

Color red.

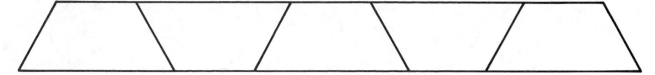

Color blue.

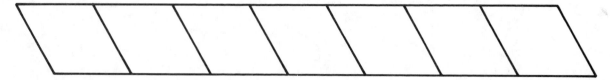

Color tan.

People Counters

Support Master 7

Teddy Bear Counters

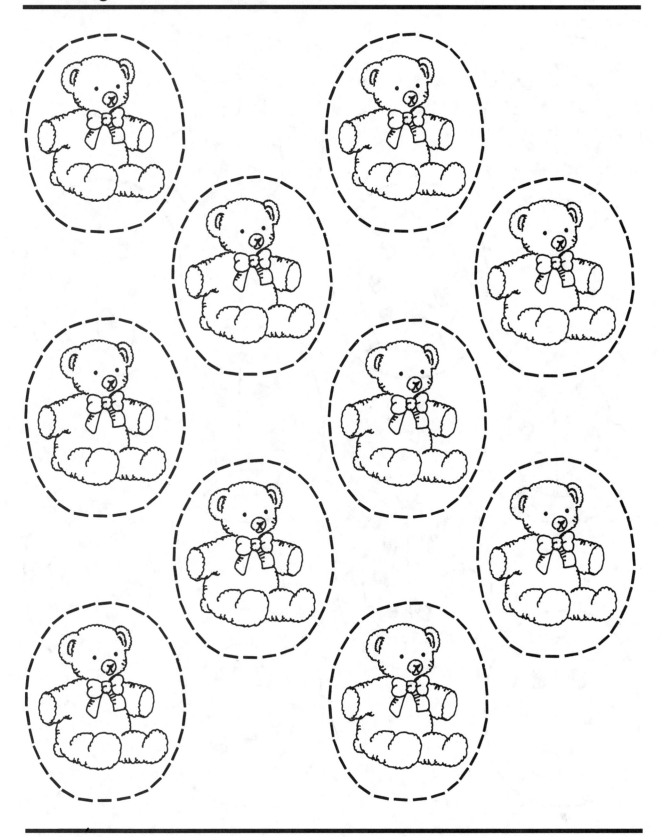

Bug Counters

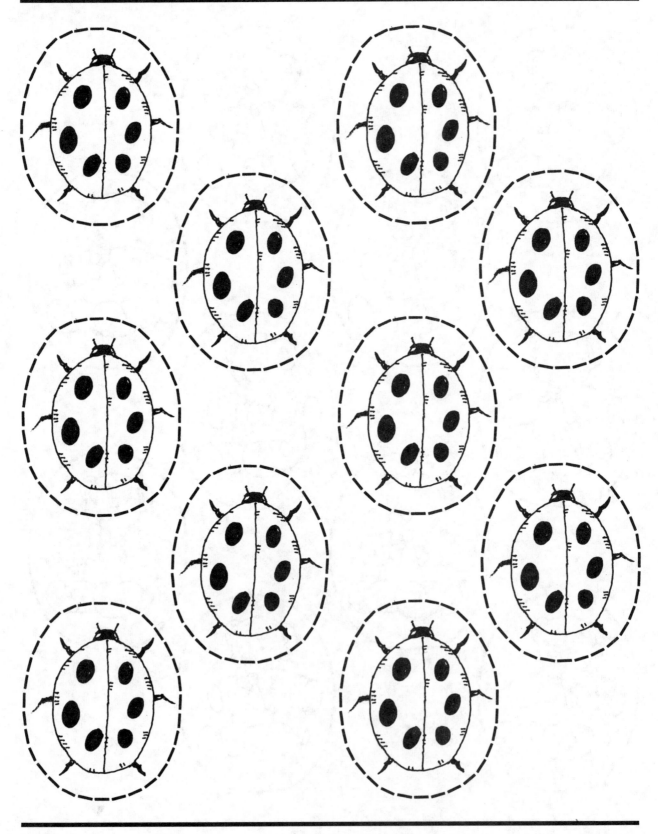

Supplementary Book List

UNIT 1
Number Sense and Numeration

- *Anno's Counting Book* by Mitsumasa Anno (Harper & Row, 1977)
- *The Most Amazing Hide and Seek Counting Book* by Robert Crowther (Penguin, 1981)

UNIT 2
Whole Number Concepts

- *Anno's Multiplying Jar* by Mitsumasa Anno (Putnam, 1983)
- *How Many Snails?* by Paul Gigant, Jr. (Greenwillow, 1988)

UNIT 3
Geometry, Patterns, and Spatial Sense

- *Changes, Changes* by Pat Hutchins (Macmillan, 1987)
- *Left or Right?* Karl Rehm and Kay Koile (Clarion, 1991)
- *The Line Up Book* by Mirisabina Russo (Penguin—Puffin, 1992)
- *Pattern* by Henry Pluckrose (Franklin Watts, 1988)

UNIT 4
Measurement, Money, and Time

- *Caps, Hats, Socks, and Mittens* by Louise Borden (Scholastic, 1989)
- *Dollars and Cents for Harriet* by Betsy and Giulio Maestro (Crown, 1984)
- *Stone Soup* illustrated by Diane Paterson (Troll Associates, 1981)

UNIT 5
Fractions, Probability, and Graphing

- *Mouse Days* by Leo Lionni. (Pantheon Books, 1980)
- *The Three Hat Day* by Laura Gerlinger (HarperCollins, 1987)

List of Publishers

Addison-Wesley
Route 128
Reading, MA 01867

Astor-Honor, Inc.
48 East 43rd Street
New York, NY 10017

Atheneum
(Division of Macmillan)
866 Third Avenue
New York, NY 10022

The Atlantic Monthly Press
19 Union Square West
New York, NY 10003

Clarion Books
215 Park Avenue South
New York, NY 10003

Dial Press
(Division of Bantam Doubleday Dell)
666 Fifth Avenue
New York, NY 10014

Dial Books for Young Readers
(Division of Penguin USA)
375 Hudson Street
New York, NY 10014

Four Winds Press
(Division of Macmillan)
866 Third Avenue
New York, NY 10022

Franklin Watts
387 Park Avenue South
New York, NY 10016

Greenwillow Books
105 Madison Avenue
New York, NY 10016

HarperCollins
(Division of Harper & Row)
10 E. 53rd Street
New York, NY 10022

Little, Brown
34 Beacon Street
Boston, MA 02108

Macmillan
866 Third Avenue
New York, NY 10022

Philomel Books
(Division of Putnam)
Box 7336
Huntington Beach, CA 92615

Scholastic
730 Broadway
New York, NY 10003

Thomas Y. Crowell, Junior Books
10 E. 53rd Street
New York, NY 10022

Viking Kestrel
(Division of Viking Penguin)
375 Hudson Street
New York, NY 10014